MISBELIEF

Also by Dan Ariely

MISBELIEF

What Makes Rational People
Believe Irrational Things

DAN ARIELY

HARPER

An Imprint of HarperCollins*Publishers*

HarperCollins books may be purchased for educational, business, or sales promotional use. For information, please email the Special Markets Department at SPsales@harpercollins.com.

FIRST EDITION

Library of Congress Cataloging-in-Publication Data has been applied for.

ISBN 978-0-06-328042-7

23 24 25 26 27 LBC 5 4 3 2 1

This book is dedicated to the misbelievers who helped me understand their worldview and, in the process, better understand the world we all share. Many of them started out as personal antagonists but became my anthropological guides. A few, in a strange way, became sort of something not exactly unlike friends. My deep thanks for your time and guidance.

Contents

DEMONIZED

An Introduction That You Should Read
Even If You Are the Kind of Person Who
Usually Skips Introductions

Never imagine yourself not to be otherwise than what it might appear to others that what you were or might have been was not otherwise than what you had been would have appeared to them to be otherwise.

—LEWIS CARROLL, *ALICE'S ADVENTURES IN WONDERLAND*

"Dan, I can't believe you have become this person. When did you get so greedy? How have you changed so much?"

I recognized the name on the email—Sharon, a woman I had met several years earlier when she asked me for help with a corporate workshop she was delivering on behavioral change. At the time, I spent three hours helping her add more substance to her presentation, free of charge. After the workshop, she called to say thanks, and that was the end of our relationship—until July 2020, when I received this strange and cryptic message.

I fired off a reply: "What exactly are you referring to?"

Her answer contained several links, and when I clicked on them, I found myself embarking on one of the most disconcerting, disturbing, yet fascinating journeys of my life. It was as if I had walked to the edge of my known reality and pulled back a curtain to reveal a parallel universe in which someone with my face, my voice, and my name was perpetrating evil deeds that threatened humanity on a global scale—and had been doing so for a while. It felt like the opening scene of a sci-fi novel. The links Sharon sent me led to numerous websites that portrayed me as "the chief consciousness engineer" of the "Covid-19 fraud" and a leader of the so-called Agenda 21 plot. In this parallel universe, my Illuminati friends and I were in cahoots with Bill Gates, designing a fiendish plan to inject women with a vaccine that would make them infertile and reduce the world's population, while simultaneously creating an international vaccine passport system that would allow those in power (allegedly Bill and the Illuminati, including me) to track the movement of everyone around the globe. Many online contributors took these ideas further and claimed that I was collaborating with multiple governments to control and manipulate their citizens.

I didn't know what to think. As I read further, I started smiling to myself. After all, it was absurd—and, in case you were wondering, blatantly false. My main connection with Bill Gates was some brief work I'd done with the Bill & Melinda Gates Foundation on early-childhood nutrition in Africa some years earlier. I've certainly never joined the ranks of the Illuminati (and even if I wanted to, I have no idea how one would do so). No Covid-19 vaccine had yet been approved and I'd played no part in the effort to develop one. As for advising governments, I did do some of that, but my advice was limited to issues such as incentivizing adherence to pandemic restrictions and mask wearing; distributing financial aid more effectively; improving motivation for educators and students; and working to decrease domestic violence. I saw myself as someone who works tirelessly to make things better, yet here were many people comparing me to Joseph Goebbels, Adolf Hitler's devoted underling and chief Nazi propagandist. Could this be anything but a bad joke? Or perhaps just a bizarre misunderstanding? Surely no one would take this seriously.

However, I followed the links deeper into the bowels of the internet, and it appeared that many people were taking it quite seriously indeed. Posts about me garnered thousands of comments. My "evil twin" appeared in clumsily edited videos, sometimes dressed in a Nazi uniform and always with despicable intentions. Online conference panels debated my character flaws and nefarious motives. There were calls for "Nuremberg 2.0 trials" to find me guilty and sentence me to public execution.

After several hours of reading and watching videos, I no longer found it remotely funny. In fact, it was painful and confusing, especially when I later learned that the people who believed these lies about me were not only strangers but included people who had previously studied and respected my work, and even some people who had known me personally for years. How could they have it all so wrong? Surely, I thought, if I could just have a conversation with them, they would realize their error and all this lunacy would end. Maybe they would even apologize.

I saw that one of the discussion leaders, Sara, had a phone number listed. She was the one calling for me to be put on trial. She was absolutely sure that once my crimes against humanity became public, I would be one of the first to be hung for everyone to watch, cheer, and celebrate. I decided to call her up and set the record straight. What could go wrong?

Well, a lot, it turned out. It didn't go well—for reasons that are probably obvious to anyone who has spent even a few minutes considering the odds of convincing someone to change their mind via a surprise phone call. Only I wasn't thinking; I was deeply offended and emotional. After introducing myself, I told her that I wanted to set the record straight and that she was welcome to ask me anything she liked. Her first set of questions surprised me. She asked me about my views on what was going on. When I started talking about Covid-19, she stopped me immediately.

"No, no, no, what do you know about how Covid-19 fits with Agenda 21 and the globalists?"

"I don't even know what Agenda 21 is," I responded, "and I am not sure which globalists you have in mind."

"Don't play innocent with me," she responded. "I know who you are and what you do."

Then she switched tracks and demanded to know what projects I was working on with different governments. At the time, I was working quite intensely on projects related to Covid-19 with the Israeli government and a bit with the British, Dutch, and Brazilian governments as well. In the face of Sara's demanding questions, I felt as if I were on trial and she was the prosecutor. I told her that I was mostly working to try to get the police to use rewards to incentivize good mask-wearing behavior and observance of social distancing instead of using fines. I was also working on how to effectively carry out distance learning in schools and trying to figure out what financial support the government should give to those who were forced to shutter their businesses.

Sara did not buy any of it. Not for a minute. "What about tearing families apart by telling grandchildren not to see their grandparents? What about increasing loneliness and stress, leading to more deaths? What about forcing kids to wear masks that decrease the supply of oxygen to their brains?"

My feeble attempts to deny all these accusations had no effect.

"How do you explain the millions you were paid for your consulting services to different governments?" she demanded. Here, in my naiveté, I saw a glimmer of hope. All the accusations thus far had been so far-fetched that I didn't know how to begin to refute them. There is a phrase in Hebrew that loosely translates to "How can you prove that your sister is not a prostitute when you don't even have a sister?" But getting paid? That I could refute. Though I do help a lot of governments, I see it as part of my academic mission, so I never charge governments for my time. Plus, like all other US citizens, I file my taxes every year and all my income sources are listed on my tax returns.

"What if I showed you my tax returns," I proposed, "and you saw that there are no payments from any governments? Would that change your mind?"

She muttered something and then abruptly asked me if she could post a recording of our conversation online. I was surprised. I'd had no idea she was recording our conversation. (Since then, I've learned that people in her "line of work" record everything.)

"No, you can't," I said.

"Are you hiding something?" she challenged me.

"No, I am not hiding anything," I responded, "but if I had known this was a public conversation, I would have prepared for it in a different way." I paused, unsure what more I could say. The conversation seemed to have reached a dead end. Eventually, after a few more false starts, I told her, "I'm sorry we could not get past our differences," and hung up.

A few minutes later, Sara published another post on Facebook in which she shared that Professor Dan Ariely—"The Professor," she now called me—had telephoned her to try to vindicate himself. But no need to worry, she told her followers, she had not let the Professor pull any tricks on her or feed her any lies. She added that the discussion had made it clear that if governments hire people like the Professor, it is only for brainwashing purposes. After all, why would governments need services from people like the Professor if what we were facing was a real pandemic? She concluded her analysis by saying, "The Professor was very insistent that he did not get paid for his services, and this insistence makes me extra suspicious that there is in fact even more going on under the surface that we will reveal one day in his public trial."

That discussion with Sara was obviously not very fruitful. You know how some people are slow learners? Count me in that camp. Even after that experience, I didn't stop trying. Next I ventured onto Telegram, the preferred social media platform of my detractors. This Russian-developed app is designed for people with low trust. The source code is posted publicly to make sure that nothing funny is taking place behind the scenes, and the platform makes it easy to record and send one-minute videos. I jumped right in and posted a series of videos replying to a slew of accusations: I was responsible for the quarantines. I was responsible for people being forced to wear masks, which damages the brain by depriving it of oxygen. I was responsible for the lack of basic human freedoms, for the fear, for tearing families apart, for asking kids not to see their grandparents, and for the loneliness of people around the globe.

Point by point, I offered rational clarifications of what I was and was not doing in my work with governments. Domestic violence reduction, yes. Quarantines, no. Motivation of kids to learn remotely, yes. Fearmongering, no. I presented evidence that clearly refuted the mask concerns. If masks deprived the brain of oxygen, wouldn't we have seen some cognitive decline in surgeons and dentists long before Covid-19? I shared my sadness about the isolation people were feeling and the impact of staying home on children. I also pointed out that I didn't agree with everything that governments were doing but that many things were complex and had lots of costs and benefits.

Each video I posted was attacked by comments and videos containing dozens of further accusations, flying across my screen like a growing swarm of angry wasps. I couldn't keep up with them, and I couldn't respond fast enough. Trying to swat them away only enraged them more. Soon it felt as if it was a thousand against one, and not one of them genuinely wanted to engage in dialogue. They took my words and twisted them into further evidence for their narrative. They threw out new claims faster than I could refute them. At a certain point, I realized that I was just providing more raw video footage for their unscrupulous editors. I gave up and deleted my videos, an action that was interpreted as another piece of evidence for my low moral character and an admission of guilt. As I logged off Telegram, I reflected that it might be impossible to reason with people who want to believe what they already believe and who already feel such intense hatred. Hate is not a conversation.

Soon the negative content spilled out of the parallel universe and into my world. My social media channels were flooded with hateful comments. People declared that they were burning my books. They called my business associates and smeared me and even my family. I began receiving death threats almost daily.

If you've experienced any form of hate—online or offline—or been a target of misrepresentation, you may have some sense of how

I felt: sometimes helpless, sometimes infuriated, sometimes scared, always wronged. I was also intrigued. Why was this happening to me? How had I become a target? Bill Gates? Okay, I get it, he's rich and famous and has a foundation that works in public health. Of course, that doesn't make him an evil mastermind, but you can see why he might be targeted. Dr. Anthony Fauci? Well, he's on TV a lot, saying unpopular things about masks and lockdowns. That doesn't make him an evil mastermind either, but again, you can see why he might draw some fire. The Illuminati? Well, if they even exist, no one really knows who they are, but they sound pretty shady, so maybe they deserve to be the target of a few conspiracy theories. But a moderately well-known social scientist who's written a few books about why people are irrational? I couldn't figure out how I'd ended up in such illustrious company.

Why Me?

I started looking more carefully at the "evidence" that was persuading so many people to hate me. The most widely shared piece, it seemed, was a video of me suggesting that in order to cut medical costs we should get ambulances to arrive more slowly, encourage smoking, and increase overall stress for the entire population. The face was indeed mine, with my half beard (if you are curious to know why I have half a beard, I'll explain in a moment), the words were mine, and I remembered giving this speech. But I had never given that exact speech. How could this be? To clarify, let me take you to Ireland in 1729.

Jonathan Swift, best known for authoring *Gulliver's Travels*, also wrote a fantastic but somewhat less popular satire that became known as "A Modest Proposal." The full name of the essay basically gives away its content: "A Modest Proposal For preventing the Children of Poor People From being a Burthen to Their Parents or Country, and For making them Beneficial to the Publick." In that essay, Swift suggests that poor Irish people might ease their economic troubles by selling their children as food to rich gentlemen and ladies. And Swift did not stop there; he got into the details: "A young healthy Child well Nursed is at a year Old a most delicious nourishing and wholesome Food, whether *Stewed, Roasted, Baked,* or *Boiled,* and I make no doubt that it will equally serve in a *Fricassie,* or a *Ragoust,*" he wrote. The essay, now often used as an example of the satirical art, makes a shocking—and effective—point about attitudes toward the poor.

How, you might wonder, is this essay related to our story? Well, in 2017, I decided to make a modest proposal of my own. I'd been invited to address a medical conference on the future of medicine. I'm obviously not a physician, so my role was to reflect on medical challenges from the perspective of behavioral economics. "The problem of modern medicine," I said, "is obviously a problem of supply and demand. People want a lot of health care, and the system can provide just so much. Now, most people approach this problem by trying to figure out how the health care system can change to provide more. I want to propose a different approach, which is for people to want less. This is clearly a much cheaper way to balance the gap between supply and demand. So how might we get people to demand less health care?" I went on to suggest, keeping as straight a face as I could manage, that slowing down ambulances might mitigate the need for expensive hospitalizations, as would increasing smoking and adding to the stress people were experiencing (it turns out that from a purely financial perspective, smoking and stress kill people who are sick faster and in total they reduce health expenses). My punch line, in case anyone had missed the joke, was "But wait, we are doing these things already."

Obviously, my intention was to emphasize that the health care system doesn't concern itself with people until they pass through its gates (an ambulance, in my story) and that we don't invest enough in preventive health, in fighting smoking, and in reducing stress. I tried very hard during that speech to stay serious, but if you looked carefully, from time to time you would see a small smile escape. Appropriately, the video on YouTube indicated that it was a humorous speech that was accepted with laughter and applause by the audience.

Unfortunately, none of that context was preserved in the hands of the video editors, who manipulated the speech to create evidence of my evil intentions. They combined excerpts with images of Nazi concentration camps and a soundtrack of evil laughter. The video ended with an ominous voiceover: "And this is the person setting our country's agenda." The text that accompanied the video made it seem that the film producers' incredible detective work had uncovered a set of facts about my true character, and they presented it as if it were an important exposé on *60 Minutes*.

Another commonly shared piece of "evidence" was a clip from a TV show I once participated in. I appeared to state that I had worked with Bill Gates on issues related to vaccinations. If you examined the clip closely, however, you would notice a slight glitch in the sentence, because the video editors took a sentence in which I talked about the project we had done with the Gates Foundation on hunger and early childhood nutrition in Africa and spliced it with a different sentence in which I had referred to vaccinations. And voilà: Professor Dan Ariely admits he is working with Bill Gates on vaccinations.

Yet another exposé-like video opened with pictures of me as a teenager in the hospital following an injury that left me with 70 percent of my body burned (that actually happened). It showed close-ups of my burned face and images of the bandages all over my body. But it took a surprising turn and claimed that my disfiguration and suffering had made me angry and hateful toward healthy people and that all I wanted was for everyone to suffer as much as I had. In reality, my injury had the opposite effect: it gave me more compassion and a drive to alleviate suffering. Again, the video ended with a dramatic statement about my role in the destruction of the world as we know it.

There were many more pieces of "evidence," some in video format, some in text, followed by derisive comments about my scars, people saying that I should have been burned to death and observations that my half beard makes me look like the devil.

Given the deliberate editing of the videos to erase context and put words into my mouth, you might expect me to conclude that there must be somebody with bad intentions behind all of it. But although the possibility of an evil adversary crossed my mind, I quickly discarded the notion. First, the edits made to the videos were not particularly high quality. Second, why would anybody have any interest in going after me? It is not that I have zero ego, but it was hard for me to imagine that I am sufficiently important for anyone to spend energy trying to take me down. My guess was that the people behind the videos were do-gooders, at least in their own minds, who had stumbled upon the unedited pieces of information, connected dots, drawn their own conclusions, trusted their own conclusions, edited the relevant pieces together to highlight the connections for other people, and then disseminated their handiwork to help other people see the light. Of course, gaining social media credit in the form of likes and comments was both an important bonus for their efforts and a motivation to continue.

I was particularly saddened by the people who said that they were burning my books (some even promised to share videos of the burnings). Those people had presumably purchased and read the books, and therefore they knew my history, my motivation, how I think, and the results of my research. How could they discard everything they knew about me in favor of a three-minute video? Even if they thought the "evidence" in the videos was worth considering, how could they weigh it against everything they knew about me and come away with so much hate and anger? There were many statements in their social media feeds about "doing their own research" and calling on others to do the same. But it was clear that no one was really doing any research beyond watching some heavily edited videos, taking them at face value, and then rushing to conclusions. The oft-quoted (and variously attributed) phrase "There is no expedient to which a man will not resort to avoid the real labor of thinking" came to mind as one that is exceptionally true in social media.

I still find myself puzzled by how I ended up being demonized in this way. But I think it is largely a function of having posted a lot of material online for people to pick and choose from; having a somewhat odd sense of humor; looking different with my scars and half beard; and working with governments on many projects. There is also an element of simple bad luck: someone started looking at me with a negative perspective, created some videos, and it became an avalanche of misinformation and hate that took on a life of its own. That was not a very satisfying answer, but it was the best I could come up with. Now I was ready to move on to bigger questions. You probably are, too. But first, a brief explanation for those who are curious about why I have half a beard.

What's the Deal with My Half Beard?

The basic reason for my unusual style of facial hair is that because of my burn scars, I have no hair on the right side of my face. Of course, I could choose to shave the other side and look less asymmetrical, which I did for many years. The somewhat more complex story behind the half beard started with a monthlong hike I took when I turned fifty, during which I didn't shave and hardly saw what I looked like. When the hike ended, I didn't like how my facial hair looked and had no plans to keep it. But it was a reminder of the trip, so I decided to postpone shaving for a few more weeks.

Then something unexpected happened: I started receiving emails and messages on social media from people thanking me for the half beard. They told me that they also had injuries and that my openness with my scars gave them a bit of courage to expose their own. Those messages brought up memories of my early days being out in the world with very visible scars. People would point at me and sometimes laugh. Parents would tell their kids, "This is what happens when you play with fire." It was pretty awful.

So I decided to keep the half beard. It did make more people look at me in a funny way and more kids laugh, but I felt that going back to shaving would be hiding my injuries instead of being clear and open about them.

Over the following months, something even more unexpected happened: the oddity of the half beard helped me feel a greater acceptance of myself. And not just of my facial scars. I have lots of other asymmetries because of my burns, and somehow having the half beard helped me change my attitude toward them, so much so that I now just think of them as a part of who I am. They simply document a chapter of my life story.

This new self-acceptance made me realize something about the daily act of standing in front of the mirror shaving that I'd done for so many years. In my case, it was not simply shaving; it was also engaging in a process to make me look less asymmetric and disguise my injury a bit. What was the impact of such daily self-concealment on the way I was thinking about myself and my scars? In retrospect, I realized that shaving/concealing was holding me back from accepting my injured self. Now that I've stopped, things have gotten much better for me.

As a social scientist who is supposed to understand human nature, I am a bit embarrassed to admit that the benefits of my half beard took me by surprise. I had not even the slightest intuition of the positive perspective change that my decision not to shave would bring about. (I also didn't have the slightest intuition that I would become known, in shadowy corners of the internet, by the Harry Potter–esque moniker "The Half-Beard Professor.") Maybe it is another reminder that our intuitions are limited and that we need to be more willing to experiment with all kinds of changes, even if we initially expect that they will not bring us any benefits.

What Should I Do?

As I spent hour after hour reading posts and watching videos about the imaginary me, I felt as though I was losing my mind. To be clear, I don't just mean that metaphorically. It was as if part of my brain was constantly engaged with the hate I was experiencing, and it left less brainpower to do my actual work. Imagine a computer that spends too much processing power on a background function. That was how I felt. Unlike a computer, however, I was fully aware that I was slower than usual. And I suspected that it would take more than a reboot to get me up to speed again. I took longer to make decisions, and I was less confident that I was making good ones. Was my IQ being eroded by my preoccupation with misinformation? Why was I unable to regain control over this part of my brain? Why was I so obsessed with their untruths that I repeatedly argued with them in my mind?

Observing my sluggishness gave me a new insight into a research topic that I had been very interested in but until then had never fully appreciated: **scarcity mindset**. Research on this topic showed that participants scored much lower on IQ tests when they were relatively poor (farmers who were a few weeks away from the harvest season, for example) compared to when they had some money (farmers who had just harvested and sold their crops). And the differences were substantial in terms of their mental ability (fluid intelligence and executive control) when they were stressed over money. My crisis was not financial, but the effects of constantly thinking about those worries felt similar. I looked at more and more research on scarcity mindset—the notion that poverty impedes cognitive capacity by taking up part of the brain's limited bandwidth—and I began to understand this effect in a deeper way and feel greater empathy for the people who were described in those papers. Being haunted by worries day and night is a heavy burden to carry. Some degree of worry can be useful because it might make it more likely that we will pay attention and make better decisions. But nonstop worrying, worrying that eats up so much of our attention and brainpower, can't possibly be useful.

Recognizing my own reduced bandwidth and its similarity to scarcity mindset was only a small insight, but making that simple connection between self and science shifted my emotional state. My dark sense of helplessness receded a little. In its place, like a glimmer of light, was an old friend: curiosity. After all, I'm a social scientist. My life has been dedicated to shedding light on human behavior, in all its wonderful irrationality, and the starting point of my intellectual adventures is often my own experience. I might be unable to reason with this group, let alone silence them, but I could strive to understand them and the impetus for the stories they created about me. In so doing, I might become a better social scientist and in the process regain some control of the story. I resolved to see where this research might take me.

Upon hearing about my intentions, my mother became worried about my safety. She asked me to first consult some experts in social media and PR, which I did. Not surprisingly, all of them had the same advice: *Do nothing.* That's the standard advice in our age of misinformation, ever-increasing polarization, hair-trigger outrage, and democratized media: *Ignore it. Don't feed the trolls!* (That's good advice, by the way, and if I had a different psychological makeup, I might have listened.) One expert even told me that to be grouped with Bill Gates and the Illuminati by Covid deniers might even boost my reputation among the rest of the population.

I did try to stop engaging with my detractors. During the day I was very busy with research and various programs to manage the social implications of the pandemic, so it was easy to stay off social media and focus on work. At night, it was a different matter. The nightmares kept coming, terrifying dreams in which I was hunted and haunted. I also had a recurring dream in which I was traveling the world, going from city to city looking for a place with less hate and anger to call home. After a few weeks of that, I realized that I could not go on that way. Amid the hurt and confusion, my curiosity grew. My coping mechanism of choice was to try to understand the phenomenon and use everything I knew from social science to make sense of what I was experiencing. I had never imagined being turned into a villain by tens of thousands of people, but when it happened, it felt as if my sanity depended on figuring out how—and why. Hence this book.

How I Approached This Book

The journey of this book began with my own experience but it quickly became about a phenomenon that affects all of us. It led me to venture into research areas that are new for me, such as personality, clinical psychology, and anthropology. The spread of conspiracy theories and the scourge of misinformation are challenges that reach beyond the realm of social science and exceed the scope of my expertise and the capacity of any single book. Technology, politics, economics, and more play a role in driving and accelerating these problems. With the advent of advanced AI tools such as ChatGPT and its siblings and the ongoing polarization of everything, it's hard to see from a societal and structural perspective how we might solve them anytime soon. What fascinates me—and where I see leverage for positive change—is understanding why people are so susceptible. Why do we not only believe but actively seek and spread misinformation? What is the process by which a seemingly rational person begins to entertain, adopt, and then defend irrational beliefs? Approaching these questions with empathy, rather than judgment or ridicule, is both illuminating and disconcerting.

In this book, I will use the term **misbelief** to describe the phenomenon we're exploring. Misbelief is a distorted lens through which people begin to view the world, reason about the world, and then describe the world to others. Misbelief is also a process—a kind of funnel that pulls people deeper and deeper. My goal in this book is to highlight how anyone, given the right circumstances, can find themselves pulled down the funnel of misbelief. Of course, it's easiest to see this book as being about other people. But it's also a book about each of us. It's about the way we form beliefs, solidify them, defend them, and spread them. My hope is that rather than simply looking around and saying to ourselves, "How crazy are those other people?," we will start to understand—and even empathize with—the emotional needs and psychological and social forces that lead all of us to believe what we end up believing.

Social science provides us with a valuable set of tools for understanding the various elements of this process and for interrupting or mitigating it. Much of the research I present in these pages is not new. I have found myself returning to some of the cornerstones of the field in my quest to shed light on the emotional, cognitive, personality, and social elements that lead people into misbelief. This isn't surprising. After all, a propensity for misbelief is part of human nature.

In many ways, this book builds on my previous body of work, especially my research into human irrationality. After all, what could be more irrational and more human than to adopt a set of beliefs for which there is little or no evidence? And to insist on the truth of those beliefs even as they alienate us from our family and friends and cause us to live in a painful state of suspicion and mistrust?

In other ways, however, this is a very different book from any I've written before. First, it's more personal than my other books. The experiences that sparked it were challenging and emotionally difficult, and doing this research meant I had to dwell on those experiences for an extended period, further increasing my discomfort. Second, this book examines a phenomenon that is much more complex and multifaceted than anything I've researched before. In the past, the research I conducted and then described in my books was about specific topics such as procrastination, motivation in the workplace, online dating, and how we misthink about money, among others. My hypotheses were precise (at least, I'd like to think so), and the research answered questions that were, I hope, both practical and theoretically interesting. The problem I set out to understand here draws its power from multiple sources and includes numerous intersecting elements. From the start, I knew it would not yield a simple answer. My hope is that I will nevertheless be able to provide a helpful framework through which to understand the general process that people undergo on their journey from being believers to being misbelievers.

My approach, by necessity, has been a combination of personal reflection, conversation, anthropological research, and a wide-ranging review of the social science literature that can help shed light on different aspects of this topic. To write the narrative elements of this book, I relied on my memory of the events described, augmented with research and the corroboration of others where possible. Spurred on by my own experience, I spent thousands of hours examining different sources of information and misinformation; listening to and sometimes participating in online discussions; reading the academic literature; and conducting my own research (and by "research," I don't mean watching YouTube videos).

In a departure from my approach up to this point, I also engaged in ongoing conversations, even relationships, with misbelievers—the very people who were propagating hate about me online. You'll meet many of these people in the pages of this book. They started out as personal antagonists but became anthropological subjects, essential to my broader research. I tried to get to know them, to empathize with them, to understand what led them down the funnel of misbelief, and then to use the lens of social science to generalize what I learned. I have changed the names of some of these individuals and modified identifying details, including physical descriptions, nationalities, and occupations, to respect their privacy without damaging the integrity of the story. I have reconstructed conversations to the best of my ability and drawn on texts, email communications, and social media posts (sometimes summarized and sometimes translated). These conversations are not written to represent word-for-word documentation; rather, I've retold them in a way that evokes the feeling and meaning of what was said, in keeping with the essence and spirit of the interaction.

My hope is that through these stories and reflections, maybe we'll come to understand a little more about what's happening in the world we all share and discuss the ways in which we as individuals, families, and societies, might mitigate it. As the scale of the problem seems ever more overwhelming, a focus on the human element—on understanding and combating misbelief in ourselves and in others—may be the most immediately hopeful avenue for change. That doesn't mean it's simple. But there are many small actions we can take to guard against falling into misbelief ourselves, to discourage those around us from adopting false narratives, and to slow or reverse the process by which someone we know and love is succumbing to misbelief. Throughout the book, I've highlighted my ideas about how to do this under the heading "Hopefully Helpful." They include various tools and insights from social science that may prove useful as we navigate these difficult waters. I sincerely hope that these suggestions will be helpful, while also acknowledging that we all have much to learn about how to untangle the web of misbelief and misinformation that has ensnared our public and private discourse.

Perhaps the most helpful—and hopeful—place for any of us to start is with understanding and empathy. Yes, the content of the misinformation we encounter can range from laughable to strange to ludicrous to offensive and even dangerous. Some of it deserves the pejorative label "conspiracy theory." But what drives people to engage with this content may be more relatable than we'd like to admit. I tried to approach the misbelievers I met with genuine curiosity, recognizing that there is nothing to be gained by dismissing, ridiculing, or canceling people whose beliefs appear irreconcilable with my reality. This is one reason why I have chosen the term *misbelievers* rather than the judgment-laden "conspiracy theorists." I hope that this approach will help all of us better understand the people in our lives who see the world in ways that may be incomprehensible to us. And perhaps, in the process, we will also get around to questioning some of our own beliefs and the way we got to those beliefs. After all, we are each, in our own ways, misbelievers.

PART I

THE FUNNEL OF MISBELIEF

CHAPTER 1

How Could *That* Person Believe *That* Thing?

I know that most men—not only those considered clever, but even those who are very clever and capable of understanding most difficult scientific, mathematical, or philosophic problems—can very seldom discern even the simplest and most obvious truth if it be such as to oblige them to admit the falsity of conclusions they have formed, perhaps with much difficulty—conclusions of which they are proud, which they have taught to others, and on which they have built their lives.

—LEO TOLSTOY, *WHAT IS ART?* (1897)

"We talk about the weather," a friend told me with a sad smile, referring to her in-laws. Most other topics of conversation—work, health, politics, even her kids—have become dangerous ground, likely to expose the gaping ideological rifts between her and the people who had once welcomed her into the family like a daughter.

These days, it seems as though we've all gotten used to having people like that in our lives—friends, family members, or colleagues with whom we carefully restrict our conversations. Perhaps they're just casual acquaintances on social media, but they may also be people we know intimately. I'd be willing to bet that almost everyone reading this knows *someone* who has undergone a dramatic shift in their deep beliefs about health, the media, the government, the pharmaceutical industry, and more over the last few years. They may not suddenly believe that the earth is flat (though a surprising number of people do). But they may well deny the existence of Covid-19 or think it's a bioweapon. They may refuse to admit the legitimacy of the 2020 US presidential election or think that Antifa staged the storming of the Capitol. They may insist on telling the *real* story behind the assassination of John F. Kennedy, climate change, the events of 9/11, or the death of Princess Diana. Some may confidently declare that all vaccines are evil. Others think that antivaxxers are actually lizard people who came up with an ingenious plot to destroy humanity. (Okay, the last one was made up by the folks behind the ScienceSaves campaign to promote vaccines. But you get my point.)

It sometimes seems that the growing tide of misinformation and false beliefs has left no community or family unscathed. And jokes about lizard people aside, it's no longer something we laugh about. When you hear the words *conspiracy theory*, what comes to mind probably isn't tinfoil hats or little green men; it's much more serious and more personal. Anytime I mention this topic, I see pained expressions. People shake their heads and tell me about their friend, their cousin, their parents, their in-laws, their kids. The ones they're afraid to invite to parties or family events. The ones they can't talk to at all. They just can't wrap their minds around how *that* person ended up believing *those things*.

I know that feeling all too well. For me, one of the most disturbing moments in my journey into the parallel universe was a conversation with a woman I'd known since she was eight years old and thought of as almost family. Not only had she embraced the story that Covid-19 was a global conspiracy to promote evil vaccines and kill people, but she believed I was a perpetrator of it all. Even her decades-long personal relationship with me did not dissuade her, and there was nothing I could say to change her mind.

It's confusing, frustrating, painful, and even scary to suddenly feel a chasm of misbelief opening up between you and someone you love;

someone you thought was, well, just like you. But now you wonder, how did we end up living in such alternate universes? How on earth did this seemingly rational, normal person begin to adopt irrational, false narratives about reality? And why now?

I often wonder if the problem of misbelief is getting worse. It certainly appears that way, anecdotally. It seems as though conspiracy theories are propagating exponentially, fueled by the internet, the Covid-19 pandemic, political polarization, and most recently by advances in AI technology. They're no longer relegated to the fringes of society, to clumsily dramatized homemade videos and private chat rooms. Now they're confidently voiced by elected politicians, popular celebrities, and cable news anchors. And they spill over dangerously into all our lives with events such as the storming of the US Capitol on January 6, 2021, and hate crimes motivated by misinformation. But only time and research will reveal to what extent they're more common or just more visible today.

We do know that the problem of misbelief predates our current era and will likely persist for the foreseeable future. Just to get a historical sense of the tenacity of misbelief, here are some ancient examples: In the year 68 CE, some ancient Romans believed that the infamous emperor Nero had faked his death and was plotting to reclaim his throne. Over the coming years, Rome was plagued by a spate of imposters claiming to be the emperor returned. Some people believed that Queen Elizabeth I actually died as a child and was secretly replaced with a boy (why else did the queen never marry and always wear a wig?). Speaking of replacements, Paul McCartney, who is eighty-one years old as of this writing, had to work very hard in the 1960s to convince certain fans that he hadn't died and been replaced with a look-alike. You may have heard of the belief that the earth is flat, but did you know that some people believe that the earth is hollow? And then there are all the conspiracy theories that insist certain historical and current events never happened—from the Holocaust to the assassination of Martin Luther King, Jr., to the moon landing to 9/11 to the massacre at Sandy Hook Elementary School. There's even a conspiracy theory about the origin of the term *conspiracy the-*

ory (allegedly, the CIA created it to discredit people who were questioning the official story about the assassination of JFK).

It's also difficult to tell where one conspiracy theory ends and another begins. Part of the nature of conspiracy theories is the assertion of connections—hidden webs of cause and effect, secretive relationships, and alliances dedicated to shadowy causes. So it's hardly surprising that the theories themselves tend to overlap and become woven together (as seen in the belief that the Covid-19 vaccine contains a 5G chip: two conspiracy theories for the price of one!). The crop of conspiracy theories that sprang up around Covid-19 drew on themes that were established long before anyone had heard of the virus, and although they featured a new cast of villains (people like Dr. Fauci, Bill Gates, and myself), they also brought back old favorites (the Illuminati, the Deep State, and vague shadowy elites). Old and new narratives became mutually reinforcing.

Case in point: As I was working on this book, I made a trip to Toronto. My flights, both outbound and inbound, had a stopover at Denver International Airport. That was fine by me—it's a perfectly good airport, provided you time your arrival to avoid the afternoon thunderstorms that are common in the summer. I enjoyed the mountain views as we made our descent, and I found a decent place to get a meal. However, had any of my online detractors known about it, they might have found it suspicious, if not downright damning, that my travels took me to that particular spot. Coincidence? They would think not. You see, a surprising number of people believe that Denver International Airport is in fact the secret headquarters of the Illuminati. The modern incarnation of the ancient brotherhood is said to meet in underground tunnels beneath the terminals. I'm sure it would have been easy to connect the dots and conclude that I was there for secret meetings with my Illuminati buddies about our plot to reduce the world's population using vaccines. And just like that, a set of conspiracy theories that have been around since the midnineties, when Denver International Airport opened, can merge with the much more recent Covid-19 misbeliefs.

Speaking of that particular airport, the theories don't stop with the

Illuminati. Some argue that those underground tunnels are the lairs of a colony of lizard people (perhaps the same ones behind the anti-vax campaign?). Or is it aliens? Others think the tunnels are in fact a bunker that will shelter the world's elites when everything goes to hell. Or that there's a whole underground city built by the New World Order. If you think all this sounds far-fetched, you're encouraged to look for hidden clues in the airport's art collection, which includes some pretty ugly gargoyles and a rather startling blue rearing horse with glowing red eyes, said to be a demonic representation of the coming Apocalypse. Don't believe that? Why else would the artist who created the horse have died in the process? (This part is actually true—the sculptor, Luis Jiménez, met an untimely end after a section of the horse fell on him in his workshop and severed an artery. His sons had to finish the already overdue project.)

Anyway, all of this is to say that you can find conspiracy theories wherever you happen to land. And although many of the personal stories I share in this book feature misbeliefs associated with Covid-19, as I have the most firsthand experience with those, my intent is broader: namely, to shed light on the psychological building blocks of misbelief in a more general sense.

Another reason why Covid-19 is a particular focus of this book is that the pandemic created extreme conditions for helping us to understand the general problem of misbelief. When else have we seen such a combination of widespread stress and fear; social isolation and loss of support systems; confusing messaging; loss of trust in institutions; political polarization; free time to spend online; and more? All of that contributed to a situation in which a large number of people adopted new and false narratives about the world in a relatively short time span.

Such dramatic changes on a wide scale have historically been something of a rarity. Indeed, if there's one thing that's become clear to social scientists, it's that changing people's opinions and beliefs is very difficult. To demonstrate this to yourself, next time you are at a boring dinner party, ask the people at the table to share something they changed their mind about pre-Covid. My guess is that you will get crickets. It's revealing how many people have no answer to this question, or at least not an interesting answer. Be honest: How would you answer this question? And think about the people you know. How many people do you know who have changed the football team they support (or any other sports team)? How many people do you know who have changed their political affiliation as an adult? Research shows that even changes in party leadership and agenda issues have surprisingly little effect on the vast majority of people.

All of this emphasizes how unusual it is that large numbers of people significantly changed their opinions and beliefs during the first few years of the 2020s. How many? It's hard to say. But the anecdotal evidence suggests that if we take the range of people who changed their opinions—from those who now trust the World Health Organization a little less to those who now think that the Great Reset is in full swing—it's a pretty large percentage. Just think about the people in your own circle. I think it's safe to say that everyone knows someone who descended into misbelief during those few years.

Lots of forces came together during that time to create the conditions in which so many people could change their opinions. Were those conditions unique to that moment in history? No. But were they unusually widespread and simultaneous? Yes. And this is one reason that understanding that period is so important.

I certainly hope that the exact conditions of the Covid-19 pandemic, like the emperor Nero, will not return anytime soon. But still, it is important to understand the underlying conditions and the psychological building blocks that can facilitate this kind of dramatic opinion change. Of course, opinions and beliefs can also change for good, but what we've seen in this era is a lot of people changing their opinions in a way that reduces their trust in people, in society, in science, and in institutions.

A Closer Look at Misbelief

A simple definition of "misbelief" might equate it to accepting a false-hood about a particular fact. This is not the way that misbelief is used in these pages. Rather, we will think about misbelief as a perspective or psychological mindset that acts as a distorted lens through which people view the world, reason about the world, and describe the world to others. Moreover, misbelief will be explored not just as a state but as a process.

As we will discuss in chapter 2, the process of misbelief can be thought of as a funnel. When first entering this funnel, a person might just have a few niggling questions about accepted truths and established sources of information in science, health, politics, the media, and so forth. At the other extreme of the funnel, all "main-stream" sources are dismissed, and people embrace full-blown alter-native truths or conspiracy theories without a moment's hesitation. Of course, there are many steps along that road.

When we speak about misbelief, we are not just talking about *them*—the people who believe all kinds of odd things. To some de-gree, we all have the characteristics of misbelievers. Many of us don't believe everything that pharmaceutical companies claim and we look outside of conventional medicine for additional health support. Many of us have questions about the way governments and public health of-ficials approached the Covid-19 pandemic and disagree with some of their decisions. Most of us know perfectly well that media networks have built-in biases and unspoken agendas, though not necessarily nefarious ones. But in general, we approach information from a gov-ernment, scientific institution, or the media with the attitude that there's a good chance the information is true. That doesn't mean we won't check or confirm it. Skepticism is healthy, and it's wise to ask questions and even do one's own research or fact-checking, especially in an era in which misinformation is so rampant.

As people progress down the funnel of misbelief, however, they reach a point where healthy skepticism evolves into a reflexive mis-trust of anything "mainstream" and genuine open-mindedness slides into dysfunctional doubt. There's a tipping point where people are no longer simply questioning established narratives but instead embracing and becoming attached to a whole set of new beliefs that they've found in the process. At this stage, they approach information that comes from a government, scientific institution, or the media through a lens of automatic suspicion and mistrust. They'll look for ways in which it is likely false and misleading. Those who are deeply entrenched in misbelief will already be sure that it's part of an evil plot—a twisted, malevolent scheme perpetrated by evil elites. In this sense, misbelief is about both the quantity of false beliefs a person holds and a general mindset of mistrust and suspicion.

One helpful analogy is to think of misbelief as being like an autoimmune disease. A healthy immune system is on the lookout for infections or viruses that threaten the body, and it acts to protect us from disease. But sometimes the immune response becomes overactive or misguided and begins attacking the body, it's designed to protect. When an autoimmune condition becomes chronic, it can affect multiple systems and fundamentally impair our ability to function in the world. Chronic misbelief is similar. Our healthy instincts toward skepticism and independent thinking become overactive and turn against us in ways that are self-destructive and debilitating.

Misbelief Is Not a Left or Right Problem

It's easy to point the finger and blame the problem of misinformation on people with differing political positions while simultaneously viewing those who share our political persuasion as scrupulously factful. But this is far from accurate. Misbelief isn't exclusively a right or a left problem; it's a human problem.

Studies have found that both liberals and conservatives consume and spread misinformation, though not always equally, and that the extreme wings of each political party are particularly susceptible. Interestingly, if you follow the misbelieving extremes far enough, they sometimes loop around, meet in the middle, and create strange

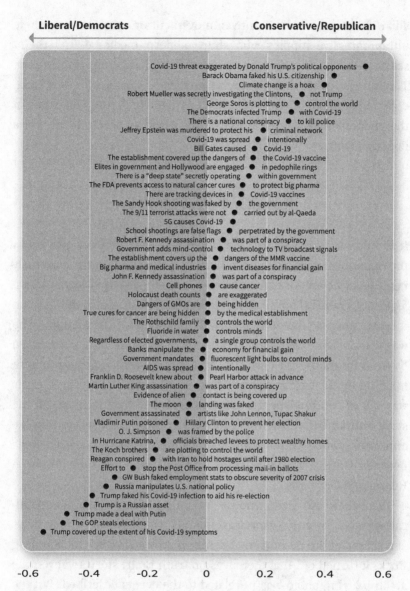

FIGURE 1

A list of misbeliefs and their correlations with political affiliation

If a belief has a "positive" score (to the right of the graph), it is more strongly held by conservatives. If a belief has a "negative" score (to the left of the graph), it is more strongly held by liberals. We tend to think that only people on the other side of the political aisle, not "our people," are likely to be misbelievers, but as this figure shows, misbelief seems to be equally distributed across the political spectrum. Based on the work of Adam Enders and colleagues.

alliances such as today's antivax movement or even QAnon, where ultraprogressive hippies who shun modern medicine find themselves aligned with ultraconservatives who mistrust big government. Though the content of particular misbeliefs changes to an extent, depending on political leanings (as illustrated in Figure 1 on page 29) the phenomenon of misbelief itself is a human conundrum, not a liberal or conservative trait.

A Medley of Misbelievers

It should be acknowledged that there is a wide range of players in the misinformation field, from nefarious to naive. On the extremely nefarious side are foreign powers who use misinformation as a strategic tool against their opponents. For example, in 2016, the Russian information machine took a story about a Russian-German girl who was missing for twenty-four hours and claimed to have been raped by Arab migrants and used it to accuse the German government of hushing up the case and concealing evidence that the refugee crisis was out of control. The girl later confessed to having made up the story but not before the misinformation had been used to provoke anti-immigration demonstrations, heightening both racial tensions between Germans and Muslims and diplomatic tensions between Russia and Germany.

Then there are those who use misinformation to promote political agendas. In 2017, American liberals widely shared a fake news story about police raiding and burning a protesters' camp at Standing Rock. It turned out that there was no truth to the story and even the accompanying photo was unrelated to the events described. But the story served to further inflame the fears of those on the left who were convinced that Donald Trump's recent election as president marked the beginning of a descent into violent authoritarianism. And of course, the same happens on the other side of the political aisle. Just one among many examples is the way Republicans spread false reports of voter fraud in order to undermine confidence in the electoral process anytime their candidates lose. Sometimes misinformation is used to cover up the consequences of misinformation being taken too far. When an attacker inspired by popular right-wing conspiracy theories broke into House Speaker Nancy Pelosi's home in 2022 and attacked her husband, Paul Pelosi, with a hammer, it took only hours for right-wing lawmakers and pundits to start spreading rumors that the attacker was in fact either a gay prostitute or a crisis actor.

Somewhat less nefarious are those who stand to gain financially from disseminating misinformation—like a health guru who makes millions of dollars selling supplements to people who are convinced that anything the pharma companies have touched is designed to kill them.

The most common type is the naive person who has no interest or agenda outside of the information itself. These people don't want to incite hate and confusion, they don't want political power, and they don't want money. They just want to understand the world around them. In this regard, these people are all of us, except that for some reason their search for understanding led them down the funnel of misbelief, and in the process they fundamentally changed their view of the world. Once that happened, they felt compelled to share their new insights and understanding of reality. At first glance, it is unclear exactly what drives them or what they stand to gain from their involvement with misinformation.

It is easy to think about these people as "them," but the reality is that they are basically like all of us. We all consume information and try to use it to understand the world around us. Sometimes we get to odd junctions, take a wrong turn, and become lost. If we want to avoid that fate for ourselves or our loved ones, it's important to acknowledge this possibility and strive for an empathetic understanding of the path that leads to that wrong turn, the psychology that underlies it, and the consequences of the journey.

CHAPTER 2

The Funnel at Work

All my friends are finding new beliefs.
This one converts to Catholicism and this one to trees.
In a highly literary and hitherto religiously-indifferent Jew
God whomps on like a genetic generator.
Paleo, Keto, Zone, South Beach, Bourbon.
Exercise regimens so extreme she merges with machine.
One man marries a woman twenty years younger
and twice in one brunch uses the word *verdant*;
another's brick-fisted belligerence gentles
into dementia, and one, after a decade of finical feints and teases
like a sandpiper at the edge of the sea,
decides to die.
Priesthoods and beasthoods, sombers and glees,
high-styled renunciations and avocations of dirt,
sobrieties, satieties, pilgrimages to the very bowels of being . . .
All my friends are finding new beliefs
and I am finding it harder and harder to keep track
of the new gods and the new loves,
and the old gods and the old loves,
and the days have daggers, and the mirrors motives,
and the planet's turning faster and faster in the blackness,
and my nights, and my doubts, and my friends,
my beautiful, credible friends.

—CHRISTIAN WIMAN, "ALL MY FRIENDS ARE FINDING NEW
BELIEFS"

When the second plane hit the World Trade Center on September 11, 2001, millions of people watched in horror as the tragedy unfolded on live television. Among them was Brad, a young man in his early twenties living in New Zealand. Unable to sleep, he stumbled downstairs, turned on the news, and was so shocked by the scene on the screen that he wondered if he was in fact still in the grip of a nightmare. In the days and weeks that followed, he couldn't stop thinking about what he'd seen: tiny figures jumping out of the burning buildings; people covered in ash running through the Manhattan streets; desperate family members seeking information about their missing loved ones. In his mind, he replayed the moment the planes hit: the fireball exploding against the clear morning sky and the sickening slow-motion tumble of the tower. A sensitive and thoughtful man, he struggled to make sense of the event and found it emotionally devastating.

A few years later, Brad's preoccupation with 9/11 continued as he traveled to the United States for work. Spending many months in an unfamiliar country without his usual support network of family and friends, he had time on his hands to think and to read. At some point, he came across a couple of documentaries about 9/11 that questioned the official accounts and offered alternative explanations for the events. Those gave him a glimpse of another possibility. He searched for more information and shared it with anyone who would listen. Not only did he find more theories about the World Trade Center attacks; his exploration led him to the work of British footballer turned social media personality David Icke, whose numerous claims include the idea that Earth has been hijacked by a sinister reptilian race. Soon Brad adopted that belief, along with various theories about UFOs, aliens, and more. Today, many years later, his central belief is that the world is run by an evil cabal of pedophiles and that 9/11 was carried out by the US government. Outside of his job as a real estate agent and spending time with his wife and two kids, he dedicates every waking hour to "doing research" and enlightening others about what's "really" going on in the world. Over the course of the past two decades, he has taken quite a journey, deep into the funnel of misbelief. He now has a whole new set of friends whom he met through his online explorations, and he's lost touch with many of the people he was once close to.

The funnel of misbelief is an amazing and complex phenomenon. People start out with one set of opinions and beliefs, enter the funnel, and come out with a very different set of opinions and beliefs. Their family and friends often watch in bemusement, unable to fathom how the person they thought they knew so well could have undergone such a shift.

Funnel Basics

In my view, the funnel of misbelief can be divided into emotional, cognitive, personality, and social elements (see Figure 2, page 36). In this book, we'll use the term *elements* because it implies multiple building blocks, each of which plays its own part in creating the structure. Of course, the distinction between the elements of misbelief is imperfect and the process is not linear, as in A + B + C + D = misbeliever. They are not four distinct stages in a process, although some elements play a more prominent role early in the funnel and others become more important later. Nor am I talking about a process that is deterministic. You could combine all the pieces I'm about to describe, and it won't guarantee that someone will become a misbeliever. However, it does make it more probable.

Our exploration begins with the emotional elements, focusing on stress, because this is the condition that sets the stage. And it ends with the social elements, because in many ways they seal the deal. As we examine each element, we'll look at its role in earlier and later stages of the overall journey down the funnel. This can be important when thinking about how to reach someone who is becoming ensnared by misbelief. For example, if emotional elements such as stress and fear appear to be predominant in their experience, it's likely that they are just beginning their journey, and there are multiple ways to reach out to them and slow or even reverse their progress.

However, if social elements, such as the desire to prove themselves or gain status among their new social groups of misbelievers, dominate, the person is likely quite far down the funnel. Extricating someone at this stage is much more difficult, though not impossible, because they've become socially entrenched in circles of misbelievers and detached from much of their old social support systems.

That being said, it's important to remember that emotional, cognitive, personality, and social elements are at play throughout the process. Imagine four different-colored liquids swirling down a funnel, occasionally blending with one another, and you will get a sense of how these elements interact. The rest of this book will dive deeper into each. I have divided the book into four parts to create some natural breaks, as each contains a lot of information. I encourage you to pause and take the time to consider and digest each part before moving on to the next. Think about how it relates both to people you know and, perhaps, to yourself. To begin, here is a short overview.

The Emotional Elements: Human beings are emotional creatures, and as social science research has repeatedly shown, emotions tend to precede beliefs. Emotions are often the main drivers of our actions. As Jonathan Haidt has persuasively argued, "Intuitions come first, strategic reasoning second." In other words, we start with an intense emotional response, and then we come up with a cognitive explanation for it. In the funnel of misbelief, the emotional elements center around stress and the need to manage it, which set up the conditions for the other elements to come into play.

The Cognitive Elements: The human mind has a tremendous capacity for reasoning, but that doesn't mean it is always rational. When we are motivated in one direction or another, confirmation bias kicks in and leads us to seek information that fills that need, regardless of its accuracy. And then the story gets more complex: we construct narratives to get to the conclusions we want to get to. And it's not just the ways we think that lead us deeper into the funnel; it's the ways we think about our own thinking that get us into trouble as we part ways with reality.

The Personality Elements: Not all human beings are equally predisposed to misbelief. Individual differences play a key role in the process by which some people descend into the funnel while others do not. Some personalities, it turns out, are more susceptible than others. Those who display certain traits have a greater propensity to embrace false narratives about the world, and although none of these personality traits is a guarantee that someone will become a misbeliever, each of them increases the likelihood of that outcome.

The Social Elements: Misbelief does not develop in a vacuum, nor is it sustained in one. Powerful social forces attract people to change their minds, lead them down specific paths, keep them among fellow misbelievers, and even accelerate the extremity of their beliefs.

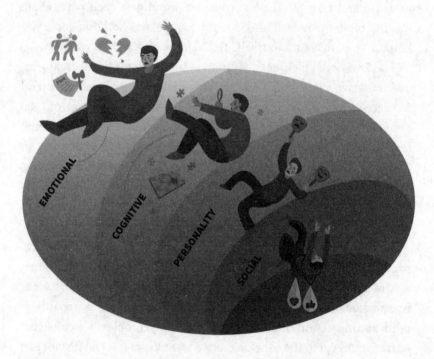

FIGURE 2

The funnel of misbelief and its constituent elements (emotional, cognitive, personality, and social)

A sense of community and belonging is a powerful draw and is particularly important in cases where people feel disconnected or ostracized from mainstream society. This is especially true when the ostracism comes from their former circles of family and friends—an all-too-common situation for misbelievers. Social networks feed the information bubble, and the social currency of likes and reactions feeds the sense of being a contributing member of the community. In the funnel of misbelief, the social elements are the components that "seal the deal" and make it exceedingly hard to escape.

The Funnel Writ Large: A Societal Spiral of Decreasing Trust

This book will tackle the psychological process of going down the funnel of misbelief primarily as a personal journey that individuals take, whether willingly or unwillingly, driven by both inner and outer forces. But if we step back and examine this journey from the broader perspective of society at large, it paints a different and more worrisome picture. The individual journey that people take down the funnel of misbelief reflects a societal journey into mistrust. No matter where you are on the political spectrum, and no matter where you are in the world (with the possible exception of Scandinavia), it is hard to escape the ways in which our society's level of trust is decreasing, with alarming consequences.

Viewing the misbelief process through the lens of trust sheds an important light on the way in which people sink deeper and deeper in terms of their focal misbelief (for example, the 2020 US presidential election was stolen, or the assassination of JFK was a CIA plot). It also helps us understand why misbeliefs attract other misbeliefs—even ones that seem unconnected. Why are those who subscribe to one misbelief likely to adopt others as well? Mistrust! The accumulation of misbeliefs makes sense when we realize that misbelief thrives on a loss of trust. When we start mistrusting one institution, it becomes easier to mistrust another. In fact, we can quickly assume that maybe all institutions are the same: corrupt, greedy, and malevolent. If the pharmaceutical companies are trying to make us sicker or even kill us, what does that say about the governments that regulate them? Perhaps they're all in cahoots, the thinking goes. And if governments are turning a blind eye to what the pharma companies are doing, maybe they are also capable of committing their own evil acts. Is it so far-fetched to think they would stage an attack on their own citizens to justify waging war in Iraq? Isn't it conceivable that a warmongering government that wanted to escalate the Vietnam conflict might have been behind the assassination of President Kennedy? In this way, A leads to B leads to C, and Covid-19 conspiracy theories lead to 9/11 conspiracy theories and JFK conspiracy theories. The common thread is mistrust.

Mistrust begets mistrust, and this is partly why the web of misinformation has so many unexpected points of connection. At the extreme end of the funnel, we find the parallel universe of QAnon, which weaves multiple threads of misbelief into one tangled tapestry. People come to QAnon from both sides of the political spectrum and find common ground in mistrusting basically everything: governments, the medical profession, nonprofits, the media, and the elites.

Though the story of misinformation and the funnel of misbelief is only one perspective on the way that trust is eroding in our society, it is a central theme of this tragic tale. It is a problem that we must understand and attempt to mitigate if we want to restore trust at a societal level. So let's begin our journey.

PART II

THE EMOTIONAL ELEMENTS AND THE STORY OF STRESS

CHAPTER 3

Pressure, Stress, Bending, and Breaking

The main thing that I learned about conspiracy theory is that conspiracy theorists actually believe in a conspiracy because that is more comforting. The truth of the world is that it is chaotic. The truth is that it is not the Jewish Banking Conspiracy or the Gray Aliens or the twelve-foot reptiloids from another dimension that are in control. The truth is far more frightening: Nobody is in control. The world is rudderless.

—ALAN MOORE, *THE MINDSCAPE OF ALAN MOORE* (2003)

Entrenched misbelief is the result of multiple elements coming together, every one of them adding their part to the total mix. As we have said, these elements include emotions, cognitive biases, personality traits, and social forces. But the starting condition and most basic ingredient that all of us can relate to is stress. To understand why stress is the condition that sets the stage for what is to come, we'll take a brief trip back to a particularly stressful time in recent history and meet Jenny, a freelancer, a single mom, and a misbeliever in the making.

Stress and the Single Mom

Remember May 2020? After the first few months of extreme Covid-related lockdowns, the warmer weather brought with it a mix of hope and apprehension. Maybe the pandemic was coming to an end. Or maybe another wave was just around the corner. Restrictions were beginning to loosen. Businesses were cautiously reopening. Like creatures emerging from hibernation, people began to venture out of the homes in which they'd been confined for the past several months. And in some places, kids went back to the classroom. Parents such as Jenny felt both relief and fear as they watched their masked sons and daughters head to school.

Jenny was happy that her son, Mike, could get back to a more normal life. Sure, she worried about the virus, but by that point she was more worried about his emotional health. The past few months had been hard on the somewhat bashful fourth grader. Deprived of any social interactions, he'd grown withdrawn and sullen, bored by long hours of Zoom classes. It had been hard on his mother too as she wrestled with the demands of supporting his homeschooling while also trying to earn a living as a freelance graphic designer serving small businesses. Her clients were also struggling to stay afloat, and her services were easy to cut. Since the start of Covid-19, her pride in being an independent freelancer had given way to envy of the people she knew who continued earning a steady salary while staying home. She needed time and space to pursue new clients, but it was hard to do with Mike sharing her home office. As she packed his lunch and got him ready to go back to school that first morning, she felt more lighthearted than she had in months.

A few hours later, however, Mike returned home in tears. When Jenny asked her son what had happened, he told her that during a break he had gone to the bathroom and in the process lost his mask. He had slipped into his seat a few rows back, trying to hide his exposed face behind a book. But the teacher's eyes had quickly landed on him.

"Mike, where's your mask?"

Mike had mumbled that he didn't have it. Looking around the room, the teacher asked the class if anyone had an extra one, but no one did. In those days, masks were in short supply. To his intense embarrassment, Mike was asked to leave the classroom. As he gathered his things and shuffled out, he felt like every eye was on him.

Listening to her son's story and feeling his pain, something inside Jenny snapped. She phoned the teacher in a rage, furious that they would single out a ten-year-old and humiliate him like that in front of his friends. The teacher had no patience for her complaints. Jenny pointed out that Covid-19 was not very dangerous for kids. Some people said that it was even less of a risk for kids than the common flu. But the teacher did not even acknowledge the logic of her argument. She was under very clear instructions from above: no mask, no class; as simple as that. All the kids had to wear a face covering to protect themselves and the other kids. The argument escalated. When Jenny hung up the phone, she was even angrier than she had been before the call. It felt like the whole world had gone crazy. Nothing made sense anymore. Her already frayed emotions were at a breaking point.

That night, after she'd comforted Mike and put him to bed, she sat down at her computer. Her head was full of questions: What's the big deal about this virus anyway? Why are they so strict with kids when it's clear that it affects primarily older people? She did not know of anyone who had died from it, despite all the stories in the news. And there were no reports of young people who had gotten sick. What's *really* going on? she wondered. And who is behind it all? The need for an explanation overwhelmed her.

Later that night, with a click of her mouse, Jenny began her life-changing descent into misbelief—a journey that would lead her to become a vocal proponent of Covid-19 conspiracy theories and a leader in the antivaccination movement. She started by looking for videos discrediting masks and she found many of them. She even found what appeared to be academic papers claiming that because the virus is so small, masks could not possibly be effective at protecting people from it. The same papers also claimed that wearing masks deprived people of much-needed oxygen and the deprivation was particularly damaging for kids during their developmental years. Soon she was convinced that the "rag" her son was being forced to wear not only failed to decrease the risk of Covid-19 but actually compromised his health. She also found assertions that if worn too long, masks could become a source of bacteria and exacerbate acne for teenagers. And a mask could force exhaled air up along the sides of the nose, drying out her son's eyes.

That was where I entered her story. As Jenny dug deeper into the mask-related content online, she realized that it was all Dan Ariely's fault! She arrived at that conclusion as a result of two facts about me. Both of those facts were true, but when connected together they created an untruth (a common recipe for misbelief). Fact number one was that I had been involved in advising the department of education. This was accurate. I had indeed advised it on several issues, primarily relating to how insights from social science could help motivate teachers and students during the difficult era of school closures. Fact number two was that as a social scientist, I had concluded that the "protect others" message was a stronger motivator than a message of self-protection when it came to persuading people to wear masks. That was also true. I'd expressed that opinion in the media and in meetings with government officials. The untruth was created by connecting those two facts and concluding that I was the person responsible for convincing the department of education to force kids to wear masks in school, leading to all the imagined health hazards that went along with the practice. In actuality, I had never talked to folks at the department of education about masks, and as much as I would like to believe that I provided some useful advice to many governments during Covid-19, in reality my impact was small at best.

To Jenny, however, the two facts added up to the theory in a clear and simple way. She recognized my name, saw that I was connected to the department of education, and connected that to the idea that I'd recommended using social forces to motivate people to wear masks. And that had negatively impacted her kid. In her mind, the mystery was solved: I was responsible for the masks. As she read more articles and watched more videos, she realized that the situation was worse than she'd thought. The damaging effects of the masks weren't an accidental side effect but a deliberate plot. They were designed to turn people into obedient sheep by limiting their oxygen intake, along with their ability to think for themselves.

Before long, Jenny came to believe that Covid-19 itself was a hoax, followed by a slew of other vaccine-related conspiracy theories. Over time, she added a few non-vaccine-related conspiracy theories. As her journey into misbelief progressed, my role gained more prominence in her mind and I became a central villain in a nefarious plot to control people globally.

How does someone like Jenny—a capable, devoted mother; a successful entrepreneur; a smart person who understands science—become a misbeliever? And why? Her story illuminates several factors that are key to understanding the emotional conditions that initiate this process. First, there is general stress, in this case the unprecedented level of stress we all experienced to varying degrees during the Covid-19 pandemic. Second, this stress creates the very difficult experience of being out of control and at the mercy of forces that we don't fully understand. Third, there is a tipping point when the desperate need for an explanation, for something to make sense of it all, leads down a dangerous path.

Stress is a powerful force in all of our lives, so when it comes to misbelief, it is important to note that the role of stress is not deterministic. Simply experiencing stress does not make someone a misbeliever. But it is one of the most important emotional elements that, along with the other elements, can increase the likelihood of someone becoming a misbeliever.

Why Is "Protect Others" Likely to Be a More Effective Message than "Protect Oneself"?

Even if you don't think that mask regulations were part of a nefarious plot, you may be wondering why I recommended that governments emphasize protecting others rather than protecting oneself in their messaging during the Covid-19 pandemic. Social science research reveals three basic reasons why the message "Wear your mask to protect others" is likely more effective than "Wear your mask to protect yourself."

First, humans have a built-in motivation for what we call **social utility**, which is basically our ability to care about others. Research shows that although self-interest has some motivating force, social utility is sometimes even more powerful. When it comes to wearing masks, reminding people about the social utility component of the action in question is likely to result in a higher overall motivation.

Second, there was the conundrum that young people were at less risk from the virus but could still spread it. For this demographic in particular, the message "Protect others" was critical because young people had less reason to be concerned about protecting themselves. When self-interest is naturally low, social utility becomes even more important.

Third, we all have a tendency to learn the wrong lessons from low-probability events. What does that mean? Let's take texting and driving as an example. Imagine that you believe the probability of your getting into an accident when texting and driving is 3 percent. (I'm inventing this number for the sake of this example. The exact risk depends on many factors.) One day, you're driving down the road and your phone vibrates. You feel a little rush of curiosity. Who is texting you? You sent your significant other an important text just an hour ago, and they didn't answer—maybe this is the reply? So you check your phone. And as luck would have it, you do not get into an accident. That's unsurprising. After all, the probability of texting and driving and getting into an accident is low. But something happens on the other side of this experience: you update your belief. Now you think to yourself, maybe the probability is less than 3 percent. Maybe it's 2.8 percent. So next time, you're even more likely to text and drive. And the next time, even more. Every time you don't have an accident, you learn the wrong lesson, and based on that you erroneously conclude that the risks of texting and driving are lower than you previously thought. It's a faulty learning cycle that occurs when we learn about low-probability events from our ongoing experience.

This type of learning happens with all kinds of low-probability events, including our estimation of our chances of catching a virus such as the one causing Covid-19. If we're exposed for only a few minutes, the risk of catching the virus is pretty low. So as we engage in occasional nonsafe behaviors and don't catch the virus, we update our beliefs to conclude that the risk is lower than we originally thought. And the cycle of lower perceived probability and increased risky behavior continues while also eroding our self-interested motivation. This is where social utility is different. When it comes to caring for others, we don't have the same faulty learning-from-low-probability loop, and as a consequence, our care for others does not erode over time. We continue to wear our masks in order to do the right thing, and we do not recalculate risk based on our prior experiences with catching or not catching Covid-19. And now back to the relationship between stress and misbelief.

Not All Stress Is Created Equal

Let's take a moment to clarify what kind of stress we are going to talk about. The very same word can be used to refer to a headache-inducing workload or a global disaster, so it's helpful to make a distinction between two broad categories of stress: predictable stress and unpredictable stress. Predictable stress includes things such as paying taxes, taking exams at school, meeting work deadlines, dealing with a tantrum-prone toddler, or spending the holidays with our in-laws. They're not fun, but they're an expected part of life and most of us manage this kind of stress quite well. Sure, we might worry when tax season rolls around and we have piles of receipts to organize and a balance to pay. We might lose some sleep finishing up those reports that are due every month. We might occasionally wonder why we became parents as we endure the daily meltdown. And we might drag our feet as we approach our in-laws' front door, carrying a pumpkin pie and dreading the inevitable arguments and lectures ahead. But none of these stresses sends us over the edge (at least, not until our father-in-law pours his third glass of scotch and starts explaining how everything has gone downhill since the sixties). For the most part, even when this kind of stress is present, we continue to make clear decisions and exercise our cognitive faculties. This is not the kind of stress that leads people into the funnel of misbelief.

Unpredictable stress, however, is a different beast. It includes things such as the sudden death of a loved one, a shocking medical diagnosis, a natural disaster, an abrupt job loss, or a financial crisis. When our beloved sister is killed in an accident; when we find out we have cancer; when a wildfire or a hurricane destroys our community and leaves us homeless; when we're laid off without warning, we experience a kind of stress that is of a different order than the predictable stress of everyday life. And the difference is not just a matter of greater intensity. These stressors are particularly challenging because they are unexpected. Of course, we know that such things *can* happen, but we don't really expect them to happen to us. When they do, and especially if they happen repeatedly, they create a feeling of helplessness. In general, humans don't cope well with unpredictable stress.

I discovered this for myself during one of the most stressful and difficult periods of my life—the three years I spent in the hospital following my accident as a teenager that left me with burns covering 70 percent of my body. I had to undergo extremely painful treatments on a daily basis. For a long time, I thought that my only challenge was the pain I experienced as a result of the injury and the treatments. But over time I came to realize that another element had made the whole experience particularly difficult to deal with: the fact that I never knew what was going to happen next. Other people were constantly making decisions for me that determined what excruciating procedure I'd have to endure in the next minute, hour, day, and week. They even decided when I would wake up, when my bandages would be changed, and how the bath treatment (a particularly agonizing but necessary procedure) would be done. That lack of control over my own treatment and destiny made those years incredibly psychologically challenging, on top of the physical pain and the injury itself. This is one reason why many hospitals have moved toward offering "patient-controlled analgesia," in which patients can dose themselves with painkillers up to a point. It turns out that even a little bit of control is extremely helpful, not just for the specific activity (pain reduction, in this case) but also for the general well-being of the patient.

A global pandemic such as Covid-19 certainly falls into the category of unpredictable stress. Sure, we might have heard that such a thing *could* happen, but we didn't really expect it to happen to us. Overnight, our lives were upended. Workplaces and schools closed their doors. As we struggled to learn more about the virus, rules and restrictions seemed to change by the day, often contradicting the previous ones. Twenty-four-hour news broadcasts became harbingers of our common mortality, announcing cumulative death counts by the hour. We worried about contracting the virus ourselves and how we would care for our loved ones if they got sick. And for many, that stress was exacerbated by economic instability, job loss, or business failure.

For some, the social isolation was unbearable. For others, forced intimacy with housemates or family took its toll. People like Jenny did triple duty as breadwinners, parents, and home educators. Spouses tested the limits of their compatibility while confined to their homes, sometimes without even being able to escape for a long walk outside. Hundreds of thousands of people experienced the wrenching loss of a friend or family member and were able to share their grief only via Zoom. And all of that took place in a climate of heightened anxiety, low trust, and escalating political polarization. Even as we closed our doors and distanced ourselves from our physical communities, we became more connected than ever to the news cycle and the cacophony of social media.

How Unpredictable Stress Leads to Learned Helplessness

What are the emotional effects of being at the mercy of unpredictable stress that we are unable to control? Some important insights into this question can be found in a series of experiments conducted in the 1960s and 1970s by Martin Seligman and Steven Maier. They wanted to study what happens to our problem-solving and decision-making capacities when we experience forms of uncontrollable stress. Since it's harder to experiment on humans, they turned instead to four-legged subjects. Warning: the following may be upsetting to animal-loving readers. These studies were performed in a less enlightened era.

Seligman and Maier conducted a series of experiments during which they subjected dogs to stress in the form of electrical shocks. To understand how the experiments worked, imagine one of those unfortunate dogs—let's call him Charlie. Charlie is a fairly small mongrel. In the first part of the experiment, Charlie is placed into a hammocklike harness that doesn't allow him to escape. There are panels on either side of his head. Suddenly Charlie feels an electric shock on his back paws. In pain and confusion, he moves around, trying to escape the shock. After a few electrical shocks and a few attempts to escape, he accidentally presses his nose against one of the panels, and the shock immediately stops. The next time, he does the same thing. Quickly, Charlie (who's a pretty sharp canine) learns that he can control the unpleasant experience of the shock.

Now imagine a second dog. We'll call her Loren. She is another small, unfortunate mongrel. Loren is also placed in a harness and subjected to a shock, but no matter what she does, she cannot stop the painful experience until the allotted time is up. After repeated cycles like this, Loren learns that she is at the mercy of the shock and cannot control it.

Twenty-four hours later, Charlie and Loren are taken to a new apparatus called a shuttle box. It is a box divided into two chambers, separated by a low barrier. One chamber has an electrified floor that can administer a shock; the other does not. When the dog feels the shock, it can jump over the barrier to escape it, and the shock immediately stops. When Charlie is placed into the box and the shock is activated, he quickly learns how to escape, jumping into the other side of the box. When Loren's turn comes, however, she feels the shock and doesn't move. Eventually she lies down in the box, whimpering. What's going on here? According to Seligman and Maier, the difference between the two dogs can be explained by a phenomenon known as **learned helplessness**. In the first experiment with the harness, Charlie (*C* for control) learned that he could control the shock; Loren (*L* for learned helplessness) learned that she was powerless to stop it. When placed in the shuttle box, Charlie looked for a new way to escape. Loren, on the other hand, didn't even try. She could have jumped the barrier, but she assumed that she was helpless. She seemed to have lost all motivation to look for a way out, even though one was readily available.

In Seligman and Maier's experiment, two-thirds of the dogs like Loren who had received the inescapable shock in the harness failed to escape in the shuttle box, demonstrating learned helplessness. This contrasted with the dogs like Charlie who had received escapable shocks in the harness, as well as a group of dogs who'd received no prior shocks; 90 percent of them figured out how to escape in the shuttle box.

The researchers later performed similar experiments on rats, with similar results. They also conducted less invasive experiments with human participants involving unpleasant sounds and even unsolvable puzzles rather than electrical shocks. Based on the results, they proposed that the experience of being unable to control a stressful situation produces three "deficits": motivational, cognitive, and emotional. In other words, when we experience repeated stress that we cannot control, it makes us feel less inclined to take action and less able to figure out solutions. It makes us feel worse about all of it. For these reasons, learned helplessness has been linked to an increased risk of depression.

In the midst of the Covid-19 pandemic, you may have noticed some of the hallmarks of learned helplessness in your own experience or in the behavior of the people around you. At the time lots of people reported feeling tired, defeated, helpless, and lacking motivation. This shouldn't be surprising when we consider that we all lived with unpredictable stress and the feeling of being out of control for months, even years, on end. The resulting motivational, cognitive, and emotional deficits may all play a role in pushing some people down the funnel of misbelief.

As if the damaging effects of unpredictable stress were not enough, the story of stress has additional problematic elements. One of these is that stress adds up across unrelated domains of life. Let's take a look at a fascinating study that reveals this.

The Cumulative Nature of Stress

As Covid-19-conspiracy theories spread around the globe, a group of researchers became curious about what drove the surge and specifically the role that stress played. Since pandemic-related stress was happening everywhere, it didn't offer much insight into why some people adopt false and irrational narratives while others do not. The researchers needed to examine another kind of stress that varied in intensity across different populations during the pandemic and see whether its magnitude was linked to a greater likelihood of belief in Covid-19 conspiracy theories.

Imagine you are living in a country affected by the Covid-19 pandemic, which could be just about any country in the world in the early 2020s. You're experiencing all the stresses described above: health concerns or even sickness; economic hardship and uncertainty; isolation; restriction of movement; and so on. I'll assume that you consider yourself a reasonable, intelligent person. So let me ask you this: How likely do you think you are to start believing in one of the run-of-the-mill conspiracy theories, such as the claim that the virus is a hoax or that vaccines contain tracking devices? My guess is that it's hard to imagine that you could hold such beliefs.

Now let's add some elements to our imaginary picture. The country in which you live is not only undergoing a pandemic, it's also in the grip of violent conflict. Perhaps there's a drawn-out civil war or a border dispute with a neighboring state. Along with the stress of Covid-19, you experience the threat of violence and destruction to your community. You hear sirens night after night. You have a make-shift bomb shelter in your basement. A packed bag sits ready by your front door. You've lost friends or family members to the conflict. You may even have been injured yourself. You have loved ones who are fighting on the front lines and risking their lives every day. Try to simulate the feeling of all this complexity and consider whether the added stress would make you more likely to believe those conspiracy theories about the virus. Again, my guess is that your answer would be no. After all, what difference does the stress of a war make to the reality of a virus? They're two entirely separate events. Sure, both are stressful and both are frightening, but they are independent of each other. Yet it turns out that there is more of a relationship than we might expect.

This is what Shira Hebel-Sela and her colleagues set out to study. They wanted to know whether being immersed in a conflict exacer-bates the susceptibility to Covid-19-related misbeliefs. Their research reveals that the answer is yes. After conducting an analysis across sixty-six countries, they were able to show that people living in re-gions with a higher intensity of conflicts were more likely to believe conspiracy theories regarding Covid-19. Why is this the case? Be-cause stress that comes from any source, including the uncertainty of living in a country in conflict, adds to the feeling of not having con-trol. Stress makes us feel that we are not in charge and that we don't know what will come next. These feelings are extremely uncomfort-able and bring with them the understandable need to mitigate this unpleasant state.

These results give us an important clue to the relationship between stress and misbelief. Stress is cumulative, and the research by Hebel-Sela and her colleagues made it clear that the source of stress does not have to be directly related to the content of the misbelief for it to shape a person's thinking about that topic. In other words, the stress did not have to come from the pandemic for it to shape a person's thinking about the pandemic. Interestingly, the study also showed that in the United States, belief in conspiracy theories is relatively higher than the country's conflict level would lead one to expect. My guess is that this is due to the intensity of ideological conflict and political polarization within the country, which add a level of instability that, from the perspective of stress, is similar to the effects of violence.

How We Misattribute Emotions

One of the reasons stress accumulates is that we are not good at identifying where it is coming from. In fact, more generally, we are not good at identifying where emotions are coming from. This is called **misattribution of emotions**, and it is the subject of one of the all-time best studies in social science, an experiment carried out in 1974 by Donald Dutton and Arthur Aron. In their study, male participants were approached by a female experimenter and asked to respond to a few questions, including writing a short dramatic story based on a picture of a woman. Once they finished their task, the female experimenter tore off the corner of a sheet of paper, wrote her phone number on it, and told the participants that if they were interested in the results of the study, they could call her later. Of course, nobody was really interested in the results of the study (it was designed to be uninteresting), so whether or not they called was a measure of how attracted the male participants were to the female experimenter. Later, their stories were rated for sexual content, which provided another measure of their sexual attraction. The key manipulation in this experiment was that for the control group, the female experimenter met them while they were crossing a solid bridge. Participants in the other group, the experimental group, were met in the middle of a relatively frightening, old, 450-foot-long suspension bridge. Dutton and Aron expected the participants on the suspension bridge to be much more anxious, and they were interested in exploring whether they would misattribute their anxiety to sexual arousal and as a consequence come up with more sexual stories and be more likely to call the female experimenter after the study. They were proven right; the participants on the suspension bridge were much more likely to feel sexually aroused and to find the experimenter attractive.

This study confirmed that when we feel something, we don't necessarily know where the feelings are coming from and under certain circumstances we can attribute the feelings to different things—sometimes incorrectly. In this example, the study showed that we can take something negative (fear) and reinterpret it as something positive (sexual attraction).

Understanding the tendency to misattribute emotions can be helpful in several ways. First, it's helpful just to be aware that when we experience stress—especially an accumulated burden of stress from multiple sources—we might pin it on the wrong thing, leading us down a misguided path in our attempts to relieve it. Questioning our own attributions and conclusions is critical. Secondly, we can use misattribution of emotions to our advantage. For example, it can be helpful to try to reinterpret negative unpredictable stress as something that is less negative and more predictable. Let's go back to Jenny's story to see how this might work. Her feelings of stress came from a multitude of sources, including the pandemic, her work, her kid's school, and so on. But focusing on the factors that were more predictable, such as her work, and attributing her feelings to those issues might have given her a greater sense of control. Instead, she focused on the pandemic restrictions and blamed those for everything, until she spiraled into misbelief.

Reinterpreting something negative as something positive is more challenging but also possible. For example, let's say we feel a lot of anxiety about the unknown future. We can attempt to reframe that future as a realm of possibilities and attribute the anxiety to excitement. It is important to remember that the misattribution of emotions happens when we first encounter a new situation. Once an interpretation is made, it is hard to change. For example, if the participants in the Dutton and Aron original experiment were to meet the same female experimenter again, their original interpretation would likely carry over and they'd continue to feel a similar level of attraction to her. This means that this reinterpretation strategy is particularly important in the early days of new stressors in our environment.

HOPEFULLY HELPFUL

Reattribute Stress

Reattributing emotions is something you can try for yourself. It's also something you might try in conversation with someone in your life who is flirting with misbelief. Let's say your friend Tim is showing signs of stress, and he's attributing it to broad, catastrophic narratives about the world that you worry might lead him to embrace destructive falsehoods. You've been keeping an eye on his social media posts, and they have only added to your concerns. You also know that Tim recently got divorced, his kids are away at college, and he's feeling fearful about his future. In talking with him, you might help him connect his feelings to these more specific events in his life rather than attributing them to shadowy forces and nefarious plots. Perhaps you can even help him reframe his life changes as opportunities for freedom and possibilities for self-discovery rather than as disruptive and unexpected events.

How Does Stress Impact Cognitive Function?

It's not easy for researchers to design experiments that study something like stress without subjecting people (or unsuspecting dogs) to unpleasant experiences. This is why much of today's research takes a different approach, studying populations that are already experiencing stressful life circumstances, like the violent conflicts in the research described above. Another form of stress that is easy for researchers to study is financial hardship. Remember the ideas mentioned in the introduction about scarcity mindset—the way in which a lack of resources negatively impacts cognitive functioning? Let's return to that research and explore in more detail how it might help explain what drives people to adopt irrational beliefs or conspiracy theories.

Why am I connecting scarcity mindset and misbelief? You might think that I'm trying to build a case for a relationship between economic hardship and the tendency to adopt misinformed views of the world. But although there is some evidence for a correlation between economic inequality and belief in conspiracy theories, that is not the primary point I am making. Not all misbelievers are living in poverty; many are financially comfortable. But as we saw from the research about societies experiencing both Covid-19 and violent conflict, stress is cumulative and generalizable, and so is scarcity mindset. What makes scarcity most interesting—and relevant to our quest to understand misbelief—is that it is another form of stress that reduces our capacity to reason, think, plan, and generally make good decisions. And it's a form of stress that can be relatively easily studied.

Though much of the scarcity research focuses on financial scarcity and certainly provides important and powerful data for those working to alleviate conditions like poverty, the same devastating effects of scarcity can also arise from challenges with resources such as time, multitasking, pain, food, medical conditions, and others. Scarcity mindset is in essence a reduction of freely available mental resources because they are being involuntarily used for another task. Therefore, a deeper dive into scarcity mindset can help us understand how we think and make decisions under the ongoing pressure that comes with a high mental stress load.

Let's take a closer look at some of the research on scarcity mindset, beginning with a particularly illuminating study that was conducted by Anandi Mani, Sendhil Mullainathan, Eldar Shafir, and Jiaying Zhao.

The scarcity mindset research team began their journey by conducting a series of small-scale experiments with volunteer participants at a shopping mall that showed how scarcity reduces what social scientists generally refer to as **cognitive bandwidth**—the general capacity for thinking that a person can utilize. However, they were not satisfied with that limited setup. They wanted to test their theories in the real world, with people who were genuinely experiencing scarcity in their everyday lives, not just artificially simulating it for an hour or two. They needed a population that was living with substantial scarcity.

But they also needed something else: a control group, a group that they could measure and compare against those experiencing substantial scarcity. One approach could have been to compare poor people with rich people. The problem with such an approach is that there are too many other differing factors between two such groups, and these factors could account for the differences in the mindsets of rich and poor. Because of these many differences, the conclusions that could be drawn from such research would be limited.

A useful experimental approach, in such cases, is to look at circumstantial changes for the same individuals over time. For example, we could look for a population that was living with substantial scarcity at some times but not at others. Such an approach would allow researchers to study the same people under different conditions and see how their mindsets changed in times of increased scarcity compared to times of relative abundance.

The scarcity research team took that approach, and they identified the conditions that they were looking for in the sugarcane fields of India. Sugarcane farmers, like most other farmers, experience dramatically different financial conditions at different stages of the crop-growing cycle. They have some months when cash is relatively plentiful and other months when their resources are stretched thin, making this an ideal setup to study how the same individuals think, behave, and make decisions when they are feeling relatively rich and when they are feeling relatively poor.

The research team used two simple exercises designed to measure two key cognitive capacities known as fluid intelligence and executive control. They repeated the exercises with the farmers some time before and just after harvest. In other words, when the farmers were feeling poor and when they were feeling more financially comfortable. The results were striking. Preharvest, the farmers got about 25 percent lower scores on the Raven's Progressive Matrices test, which is a nonverbal test used to measure general human intelligence and abstract reasoning. Preharvest, they were also about 10 percent slower in terms of their executive control and made 15 percent more mistakes. They were the same people! There were no differences in their personality or real IQ preharvest vs. postharvest.

The same dynamics are at play in other stressful situations that have nothing to do with poverty. Have you ever had the experience of being so worried about finishing a project on time that you just can't think clearly? Or made poor buying decisions because you were preoccupied with something really bad you thought was about to take place at work? In these and countless other scenarios, stress reduces our mental bandwidth and promotes short-term thinking, causing us to grasp for quick-and-easy answers that may not be accurate. In my case, the time I was vilified online was not just unpleasant and sometimes scary; I also noticed my cognitive functions becoming impaired under the stress of being constantly attacked. I began to make poor, shortsighted decisions. As I write this, more than two years after this odd adventure began, it is hard for me to recreate the experience of stress and scarcity mindset I felt at the time, much in the same way that it is hard for me to fully remember the pain I experienced in the hospital (and I am very grateful for that). The depth of scarcity mindset is not something we can easily empathize with unless we are experiencing its full intensity in the moment. Like the farmers after harvest, when we're free from the immediate emotional burden, we regain our cognitive function and forget much of how we felt a few months earlier.

Speaking of pain and scarcity, research has found that economic insecurity not only creates emotional stress; it can actually increase the intensity of physical pain. In several studies, Eileen Chou, Bidhan Parmar, and Adam Galinsky showed that when economic insecurity increases (either in the specific form of unemployment or in a more generalized form such as concern about our country's economic health), the experience of physical pain escalates, both in terms of self-reported pain and as measured by an increased consumption of over-the-counter painkillers. Stress quite literally hurts. And when we are hurting, we think and act in ways that are even less beneficial.

Resilience: The Forces That Help Combat Scarcity

A seemingly obvious solution to the problem of scarcity-induced **bandwidth overload**, as researchers call it, could be to remove or reduce the amount of stress a person is under. But that's easier said than done. In most cases it's impossible, especially when the stress is caused by other people or by unpredictable events like a global pandemic. If we can't remove stress, what can we do about it? One approach is to become more resilient to stress and scarcity mindset. As I tried to understand what pushed some people into misbelief but not others, the issue of resilience stood out as a key ingredient. If we're all experiencing stress, why do some of us handle it better than others? What makes some people able to think clearly, keep their eyes on the long term, and make more reasonable decisions, while others become myopic and cling to misbelief in order to deal with stress and scarcity mindset?

Resilience is a very complex construct that draws its strength from many sources. From everything we know about resilience, it seems to work as an insurance policy against stress, helping us deal with the difficult moments in our lives.

One of the most fundamental ways to understand resilience is based on the psychological construct known as **secure attachment**. Here's a story that explains the basic idea. Imagine that you're the parent of a four-year-old kid—we'll call her Neta. One day, you take Neta to the playground. And you sit on the bench, because, after all, you are no longer four years old. "Go ahead and run to the swings," you tell Neta. Neta runs to the swings, plays for about fifteen minutes, and then comes back to you. If this happens, you have managed to create a kid with secure attachment. Congratulations. Now compare this to another kid—we'll call him Amit. You tell Amit, "Run to the swings." He starts walking, and every minute or so he turns around to see if you're still there waiting for him. If this is the situation, you have not managed to raise a kid with secure attachment. Of course, these are clear-cut examples of two extremes, and there are many degrees of secure attachment in between.

Secure attachment is formed in childhood; it basically allows us to go through life knowing that if something bad happens, somebody will catch us. We don't have to look around all the time, wondering if somebody is there for us or not. When we have a very high level of secure attachment, it's a sort of ideal insurance plan, one that we can trust to cover everything. It's an amazing, magical feeling to walk

---

HOPEFULLY HELPFUL
Improve Secure Attachment

A high level of secure attachment is quite rare and difficult to create and attain, but any improvements that can be made in this regard will surely help guard against the descent into misbelief. Secure attachment is a continuum, and even a small increase on the scale pays dividends in terms of the way we view upside, downside, risks, potential, and the actions that we're willing to take. The more securely attached we are, the more resilient we are, and the less we need alternative stories to explain the world around us.

Secure attachment begins in childhood. So if you are a parent and would prefer not to raise a misbeliever, this is something you can work on directly with your own children, ensuring that they feel you are there for them, you trust them, and they don't need to fear abandonment. In adulthood we can increase our secure attachment through forming and maintaining deep and trusting relationships that will buffer us in times of stress. We can also be that buffer for others in our lives and, in the process, protect them from the tendency to fall into misbelief.

around the world this way. It gives us confidence to do things that we would not do without that level of secure attachment. For example, if we know that somebody will catch us if things don't work out, we might start a new business. We might be willing to go out on a limb and try a new romantic adventure with somebody we think is way out of our league. We might be willing to study something that we're not sure we'll be successful at. We might be willing to move to a new city or take a chance on a new job. The list goes on and on. In short, secure attachment is the kind of feeling that makes people focus more on the upside of everything and worry less about the downside.

Increasing Resilience

One of the most interesting and important research projects on resilience undertook the challenge of improving scarcity mindset. Jon Jachimowicz and his colleagues set out to see if adding factors that increase resilience could be a partial remedy to the problems of cognitive bandwidth overload, leading to clearer thinking and better decision-making even while the conditions of scarcity persisted. Consider the following.

Imagine you're one of the approximately 10 percent of the world's population who lives on less than $2 a day. Scarcity is your everyday experience. You can barely feed yourself and your family. Any luxuries, even small ones, are out of reach. Now imagine that one day, a generous person shows up in your village and offers you a choice: They would give you $6 today, or, if you were willing to wait three months, they would give you $18. Which would you choose? Anyone who can do basic math would see that getting three times as much money three months later is the better choice. Yet in conditions of extreme poverty, people often, understandably, choose the immediate, lower option, showing one more way in which scarcity affects our critical ability to delay gratification in favor of a larger reward. Predatory loans, such as payday loans, are one example of how this creates devastating effects on the financial lives of the poor. The stress of poverty compels people to take on loans with punishingly high interest rates that provide some short-term relief but push them deeper and deeper into debt.

Now ask yourself what might help you make the wiser choice in the scenario described above? What if we added a few resilience-related components to the story? What if you had a neighbor who'd be willing to share their dinner for a day or two? Or what if your community included someone you could turn to for help and advice? Knowing this, would you be able to think through your options differently and delay immediate gratification for a larger reward later? This is the question that Jachimowicz and his colleagues set out to study. Their hypothesis was simple: people who felt a greater degree of trust in and support from their communities (a kind of secure attachment provided by the community) would be less likely to suffer from scarcity mindset, even when experiencing the same complex economic circumstances. Consequently, they wouldn't need to make as many desperate decisions. In other words, a community could provide resilience, which in turn would create a buffer against the harmful effects of stress.

To test this hypothesis, they undertook a two-year study with people living below the poverty line in Bangladesh. They gave half the population access to trained volunteer intermediaries who would help them during times of need in their interactions with local government councils as they tried to gain access to public services. The other half of the population had no intermediaries to turn to in times of need. After two years, the researchers compared the decision-making of the people who had someone to turn to for help with the decision-making of those who didn't. Using the same offer described above ($6 now or $18 in three months), they found that the people who'd received the resilience intervention, and therefore had higher levels of community trust, were more likely to forgo the short-term smaller gain for a greater reward in the longer term. They also conducted related research in the United States and found that people who had greater trust in their communities were less likely to take out payday loans.

HOPEFULLY HELPFUL

Support a Misbeliever

It is difficult to eliminate stress entirely, but it is important to reduce it, particularly in the early stages of the funnel, before people become fully committed to an alternative narrative. Providing reassurance to someone in stressful circumstances can make a big difference. We may be unable to mitigate the causes of the stress, but if we can make the person feel seen and supported, it goes a long way toward slowing the spiral of grasping for answers and control. One specific approach to stress reduction that we tested in my lab at Duke University was to make people feel that they are smart and successful. We found that after such reassurance, they were less likely to go down the funnel of misbelief. This type of reassurance has a calming effect by letting people know that they are emotionally safe and are not under attack or underperforming. It can short-circuit the search for a villain to blame or a story to explain. Our experiments focused on in-person interactions, but even in online relationships reassurance may play a more positive role than we would expect. In general, humans tend to underestimate the effectiveness of emotional support and fail to appreciate how much even the smallest reassurances matter.

Offering support and reassurance sounds like a simple action, but if you're dealing with a misbeliever who is expressing alarming ideas and saying things that you think are outlandish, it's not a simple task. And when we are under stress ourselves, it can be easy to tend to our own needs first—in line with the old "Put on your own oxygen mask first" advice. But in this case, that's a mistake. The early stages of stress are the time when we can have the most impact on a person who is in the process of moving into misbelief. This is why we need to take stress more seriously both in ourselves and in our loved ones. Once someone progresses further into the funnel and their cognitive processes and social bonds reinforce their misbelief, it is much harder to reach them.

The Fraying Effects of Economic Inequality

If community cohesiveness is so important for resilience, what factors might contribute to or erode community cohesiveness and build up or destroy resilience? Jon Jachimowicz and his colleagues, including several of the researchers who'd conducted the study in Bangladesh, set out to examine an important force that they suspected might "fray the community buffer," as they put it: income inequality. Why would economic inequality have such an effect? Possibly because as economic inequality increases, people might feel less connected to their community and more alone in their struggle for economic stability. The researchers' hypothesis was that inequality would fray the community buffer for both relatively rich societies and relatively poor societies, as long as the level of inequality in a society was high. By studying communities in the United States, Australia, and rural Uganda, they found that higher levels of economic inequality indeed decreased reliance on the community for support and created greater financial pressure.

What about the different levels of wealth? The researchers found that all societies, regardless of their affluence level, suffer socially as income inequality increases, but, as you might expect, the suffering was unequal across wealth levels. The negative effects of inequality on those living in relatively well-to-do neighborhoods (and yes, even very expensive neighborhoods can have a high level of economic in-equality) were not too bad because those people were financially se-cure and had enough resources to manage economic ups and downs on their own. For those living in relatively poor neighborhoods with high levels of income inequality, the effect of having no community to turn to was devastating.

The lessons from this research are very straightforward. First, it should be clear that communities are an important resource that builds resilience. Second, a lot has been said about the ills of eco-nomic inequality and the unfairness of wealth distribution, but in addition to the fairness issues, economic inequality has the direct negative effect of reducing social belongingness in a community while weakening people's resilience. Finally, people in a lower social economic status who are more likely to need help are also harmed the most by communities with high economic inequality.

What Can We Learn from Scarcity Research to Mitigate Misbelief?

Cash flow–challenged sugarcane farmers in India and low-income villagers in Bangladesh and rural Uganda may seem a world away from our everyday life and from the people we know who have embraced misbelief. But we can take away several key pieces of information from these studies to help us understand how stress generally impacts human cognitive functioning and decision-making and how that applies to misbelievers. To summarize: Stressful conditions tax our cognitive bandwidth, reducing our ability to think clearly and exercise executive control. Stress also hurts our ability to make rational long-term decisions that require delayed gratification. Living in a community in which we feel a sense of trust and support acts as a buffer against the detrimental impact of scarcity. However, a higher level of income inequality in our community can fray our sense of social trust.

Think about how these insights apply to the stressful circumstances we endured during the Covid-19 pandemic. The pandemic makes for an interesting case study on the role of stress in our lives because unlike, say, a natural disaster, or a tragedy within a particular family or community, the stress impacted all of us. The threat of the virus itself was a common thread connecting factory workers in China to the queen of England; Hollywood A-listers to small-town schoolteachers. Nobody was fully immune to the uncertainty, confusion, and fear that accompanied the crisis. No matter where in the world we lived, no matter what our socioeconomic circumstances might have been, those were unusually stressful times, though, of course, not equally so for all.

The pandemic also shone a spotlight on the issue of community trust for many people. Some felt buffered by the support from their employers, friends, neighbors, family, even the government. Others felt deeply alone, with nowhere to turn for help. Income inequality also loomed large for many. As I've often heard it said in recent years, we're not all in the same boat, even though we're sailing on the same stormy seas.

The Compounding Effect of Feeling Hard Done By

As people looked around and saw other people's better boats, it's unsurprising that some felt the effects of the pandemic-induced stress more acutely than others. Again, this is not just about money. In some cases, our neighbors' boats may be bigger and more luxurious. But they may also appear to be supported by a bigger crew, be more stable, or be better equipped to handle the conditions of the moment. It's human nature to compare our lot with that of others, to covet our neighbors' boats, so to speak. But for some people, this comparison can easily turn into a more corrosive emotion—the feeling of being hard done by. Not only do they feel at a disadvantage; they also feel that they are in some sense being singled out for a harder lot in life, subjected to additional or unfair hardship or persecution.

This feeling of being hard done by is prevalent among the misbelievers I've met, and I think it goes a long way toward explaining the recurring theme of "the elites" that we find in many common conspiracy theories. If you're convinced that you're at a disadvantage, someone else must have an (unfair) advantage. If you're suffering unusual hardship, someone else is getting off easy. If you're lacking control, someone else must be controlling things. The vague and shadowy "elites" are a convenient repository for all of these resentments. Why are they so appealing? Well, if anyone has the power to make our lives miserable, it is most likely them. (Who else could it be?) They are distant from the rest of us and somewhat unfamiliar. (How many people do we really know who belong to the elites? What do they have for dinner? How do they manage their relationships?)

At the same time, they are familiar and we have common knowl-edge about them (we know some of them by name or picture and have learned a few details about their life and work). All of these fac-tors together make "the elites" a very satisfying target for conspiracy theories.

Of course, if you have ever spent time with these so-called elites, you might quickly discover the biggest flaw in the assumption that they're controlling the world: namely, that they are often not terribly good at organizing anything. I've had the dubious honor of attending the World Economic Forum at Davos multiple times. It's a bastion of the elite and the target of numerous conspiracy theories. I'm sure that burnished my villain credentials for many. What I learned from those meetings is that these folks can't even put on a conference with interesting content, much less orchestrate a vast and complex con-spiracy to deceive and manipulate billions of people. One point on which I fully agree with the misbelievers is that observing their ex-travagant parties from the outside can make anyone feel hard done by in comparison.

When I think about feeling hard done by, it brings to mind a won-derful story by Douglas Adams, in his *The Hitchhiker's Guide to the Galaxy* series. It features a truck driver named Rob McKenna who seems to be unable to escape the rain. Everywhere he goes, every sin-gle day, it rains. He keeps track of his misfortune in a logbook. The nonstop rain that follows him makes him grumpy and miserable. He feels supremely hard done by—persecuted by the clouds that follow him through life. What he doesn't realize is that he is, in fact, a Rain God. He is beloved by the clouds, which just want to be near him and constantly water him. He really is being singled out, albeit out of love, not malice. Eventually, he realizes that he is the one controlling the rain, and turns his Rain God status into a lucrative career, get-ting paid to draw the clouds away from vacation spots and important events.

But I digress. Let's return for a moment to our misbelieving sin-gle mom, Jenny. Jenny definitely felt like the rain clouds of misfor-tune were targeting her. She felt hard done by because of the unique financial stress she felt as a freelancer and the injustice and social humiliation she perceived in the way her son was treated. Together, those factors created a strong desire for relief from the ambiguity and uncertainty she was feeling.

Imagine that you're living with the general experience of unpredictable, uncontrollable stress. You're feeling exhausted, worn down, and helpless—like one of those poor dogs in the shuttle box. You're losing your energy and motivation to try to improve your circumstances. It feels terrible. But wait! There's more! Now imagine that in the midst of that experience, you start feeling that you're being singled out for particularly harsh treatment. You're the only one that hasn't been given a way to escape! That's how someone like Jenny feels when she begins to tip over the edge.

It starts with the general stress of the pandemic, exacerbated by financial struggles and the ongoing difficulty of keeping her business going with her son at home. It is then compounded by Mike's unfair treatment and humiliation at the hands of his teacher. All of these push Jenny to a breaking point, and her search for answers begins. Her conviction escalates as she watches video after video online explaining that the stress she's feeling is the result of a deliberate, nefarious plot. There is no real virus. It's all a hoax intended to subdue and control the population. The masks are designed to block the flow of oxygen to the brain and turn people into sheep. *That's* why she feels so foggy and sluggish. Oh, the poor sheep-people (or sheeple, as misbelievers often say) who don't see what is going on around them! Soon the vaccine will be available, and then all the people with secure jobs, large homes, and supportive networks will be herded into vaccination clinics and implanted with microchips that will track their every move. The more she reads and the more she watches, the more she feels betrayed—by the system, by the government, by the school, and eventually by me, someone whose ideas she once respected.

HOPEFULLY HELPFUL
How to Have a Nondebate

Listen and support; don't argue. When someone comes to us with a flood of grievances and stories about who's to blame, it can feel like they're inviting a debate and want us to argue and explore the ideas they're expressing. There's a sense that they're searching for the truth. And if we disagree with their views, we may feel compelled to refute their claims. In reality, people need to feel heard, seen, validated, and understood in order to create the conditions for any kind of fruitful discussion. The first thing to understand is that the real issue is not the facts, allegations, and stories; the real issue is emotional. We can empathize with someone even if we don't agree with their interpretation of why they are suffering. Simply acknowledging where they're coming from and empathizing with their pain, without trying to correct them, goes a long way. And if we then move into a debate, we'll be starting from a much better place.

These principles have been tested by Joshua Kalla and David Broockman in an approach called *deep canvassing*. Deep canvassing starts by asking sensitive questions, listening to the answers with real interest, and then asking follow-up questions as a way to start a conversation. Why is this approach more likely to be effective? Because in the standard argumentative approach we tend to start counterarguing (at least in our heads, but sometimes out loud, too) before the other person has even finished laying out their arguments. This is obviously not a way to have a productive discussion about anything.

The crux of the "not really being in the mindset to be persuaded" problem is our almost universal need to be correct, including the need to be associated with groups that do the right thing. When our feeling of being correct is challenged, it threatens our identity, and we work hard to fend off the attack and protect what is so important to us: our feeling that we are correct.

Deep canvassing helps combat this dynamic, because it starts with a genuine interest in understanding and an open-minded attitude. When an exchange starts with this approach, some of the defense mechanisms are reduced and both sides can become a bit (just a bit; remember, this is not magic) more open-minded.

Lunch with a Misbeliever

In the summer of 2020, not long after I heard Jenny's stories and accusations, I encountered another misbeliever, Eve, whose story provided an even more illuminating example of what happens when someone who is already experiencing intense and unpredictable stress begins to feel particularly hard done by. I'd first met Eve two years earlier, when she had been volunteering with a start-up nonprofit. She had contacted me for advice on how to more effectively promote her cause. I had met her in a coffee shop and tried to be as helpful as possible, and we parted ways on friendly terms. When the misinformation about me began to spread, she called me up. After reminding me who she was, she got right down to business, listing the now-familiar litany of false accusations, all of which she clearly believed. She wasn't truly listening to my side of the story, so the conversation didn't get very far. Eventually, frustrated, she declared, "I have to get to the bottom of this. We need to meet in person."

That was early in the pandemic, when meeting in person was considered highly risky behavior. But I agreed to meet her, provided we sat outside. I even ordered us lunch. She showed up late and launched into the same string of accusations. I told her, again, that they were untrue.

"My friends warned me not to come and talk to you," she declared.

"Why?" I inquired.

"Because they said you'd put a spell on me."

I could see the logic at work. If she left the conversation convinced that I was a villain, that would be good. She would have succeeded in eluding my evil powers! But if she left thinking anything else, it was because I had put a spell on her. I was the villain no matter how our conversation went.

After a short while, I gave up trying to clarify what I was and was not involved with. In an effort to learn something from our time together, I shifted the direction of the conversation and started asking her questions instead. I was genuinely curious. Just two years before, Eve had been a confident and articulate activist, committed to a worthy cause. Now she seemed fearful, paranoid, and beaten down by life. What had happened to her in the short time since our meeting in the coffee shop? The transformation I witnessed was much more than the general pandemic-induced stress and fatigue that I saw on the average face.

Eve slowly began to tell me her story. She was an art teacher, but unlike most teachers, she was not part of the union. She'd been teaching art at the same school for more than a decade and was dedicated to her students, but her status in the educational establishment was not as secure as that of her unionized colleagues. That had never been an issue until the pandemic hit. Unlike many of the other art teachers, Eve lost her job. From there, it got worse. With no monthly paycheck, she couldn't keep up with her rent and lost her apartment. "I'm living in my car!" she told me. "I'm living in my car!" She kept repeating the sentence, as if she couldn't believe it. I also couldn't believe it. Her sense of being unfairly persecuted—singled out for hardship and humiliation—was intense.

As Eve got up to leave about an hour later, I asked her, "Did you change your mind about me in any way?"

"I'm not sure," she replied. And with that, she got into her car and drove away. Interestingly, the story was not over. A few months later, she wrote me out of the blue asking for a contribution to a crowdfunding campaign she'd launched to pay for some medical expenses. I wrote back politely explaining that I'd already exceeded my allocated funds for charitable contributions for the year. She was outraged, and her response was offensive and laced with the poisonous taint of feeling hard done by. She said that she was now convinced that I was evil and that everything people said about me was indeed true.

The Need to Seek Relief

Eve, like Jenny, was driven by multiple emotional forces: general stress; unpredictable stress; a lack of resilience; the fraying of community; feelings of being hard done by. None of these is unique to these two women or even to our particular time or circumstances. But it is easy to see how all of these elements intensified during the Covid-19 pandemic and how they affected some people much more dramatically than others.

When we can't control or influence the stresses in our environment, when we feel helpless in the face of other people or events, we are driven to seek emotional relief in other ways. What is a relatively easy way to reduce the uncomfortable feeling of being out of control? Simple: find explanations for what is happening and find someone to blame. It doesn't matter if the explanations are accurate, let alone true. We just need them to provide some degree of comfort. In the next chapter, we'll look more closely at what happens when someone begins their search for relief.

CHAPTER 4

Picking a Villain as a Way to Regain Control

Hatred has its pleasures. It is therefore often the compensation by which a frightened man reimburses himself for the miseries of Fear. The more he fears, the more he will hate.

—C. S. LEWIS, *THE SCREWTAPE LETTERS*

Humans have always loved villains. Without them, the entire canon of our storytelling would be very dull indeed, from the ancient heroic sagas to today's big-budget action movies. A good villain helps us draw bright moral lines, create drama and tension, and find a satisfying resolution when evil is defeated. Our villains aren't always fictional. History has provided plenty of real-life embodiments of evil who make their fictional counterparts look mild by comparison. These days, however, we see a troubling trend. The language that was once reserved for truly repulsive characters—Adolf Hitler, for example—can now be directed at anyone we disagree with. Spend even a few hours on social media reading about the contentious issues of the day, and you'll find dozens of people being called Nazis, fascists, authoritarians, genocidal, and so on. It's no longer enough to say someone has a different view; we now have to make them a villain. Hillary Clinton can't just be a shrewdly political liberal and feminist that others disagree with; she has to be a baby-killing pedophile mastermind. Donald Trump can't just be a brash, ambitious, populist leader that others disagree with; he has to be a fascist secret Russian agent set on destroying democracy and installing martial law. And it's not just presidential candidates who get the villain treatment; it's also regular people we encounter in person and online.

I experienced this all too personally during my own unexpected demonization. One particularly unforgettable morning, I woke up to dozens of social media notifications. At the time, I was trying to avoid such things, but the sheer volume was impossible to ignore. Something was happening that involved my name, and even though I knew it was likely to make me feel worse, I was compelled to look. The post in which I'd been tagged began with the following rather graphic and gruesome paragraph:

> If a bull being led to slaughter watches his brothers being massacred before his eyes, smells their blood, and feels death in the air he erupts in rage and fights for his life until the very last moment. But in the agricultural industry, there are animal psychology experts who know how to get that bull to go straight to his death in peace and without drama. Ideally, the bull would never know he was about to be butchered, except in the last seconds of his life while completely neutralized, hanging from his legs upside down with the slaughterer's knife glistening in front of him. Human consciousness engineering is not fundamentally different. Its purpose is to lead the people toward the final destination while unaware of the ultimate goal, seemingly of its own accord, one step at a time.

The poster was named Jon. After describing the unfortunate bull, he went on to explain in some detail how the Nazis had used the same kind of "consciousness engineering" tricks to send millions of Jews, Romani, queer people, and people with disabilities to their deaths during the Holocaust. And then it got to me. (Because naturally, one connects slaughterhouses to gas chambers to Professor Ariely.)

Like the mass murderers of animals and humans, Jon explained, I was the head of the "consciousness engineering operation" to lead unsuspecting populations to their destruction. Specifically, he accused me of forcing people to wear masks in order to convince them of the "false" epidemic; preventing people from visiting their grandparents in order to isolate them and coerce them to get vaccinated; and labeling vaccine opponents as conspiracy theorists in order to discredit their fight against the "predatory establishment." He also declared that I was using my half beard to manipulate people into seeing me as a victim and distract from my true evil intent. In closing, he called for my trial to determine "the real issue," namely, "whether life imprisonment with hard labor will suit Ariely's crimes or whether a death sentence should be sought."

The post garnered more than a thousand comments and likes. Some readers even declared that death was not a severe enough punishment for criminals like me. We were worse than the Nazis, they said. I and my fellow Illuminati should be locked in a zoo and publicly humiliated. I was described as "monstrous" and compared to a demon, a devil, and the biblical villain Haman.

I share the story of this post not to highlight the hostility against me. My point is that such things have become commonplace. But it does provide a vivid example of the level of vitriol that has become normal in such discourse aimed at whoever has been designated the villain.

As I encountered such sentiments almost on a daily basis, I found myself wondering why. Why do they go to such extremes? What is driving these misbelievers to believe that the world is run by such evil actors? I was starting to understand, as we explored in the previous chapter, how the experience of compounded, unpredictable stress pushes some people to such extremes. I saw how the feeling of being unfairly treated and hard done by created deeper mistrust and frayed social bonds. I recognized the desperate need for answers, the urge to seek control, that drives some of these people to take their first steps into the funnel. I didn't yet understand exactly what they found there. I recognized the emotional challenges and needs that led to misbelief, but what were its emotional rewards? What emotional relief did they find in such negative narratives, filled with evil intent and covert plots? Why did they keep going back for more? And why did some become mired in toxic emotions like hate and aggression toward people they'd never even know? Those are the questions we'll turn to next.

It seems to me that humans generally are not wired to view the world as governed by evil. Most people believe in a positive power in the world. Some think about a benevolent God, others believe in karma or a loving universe. And even when a devil is in the mix, the dominant force is still a good one. It makes clear psychological sense to believe that the world is generally governed by positive forces. Waking up in the morning and believing that the world is governed by forces that have our best interests in mind is a comforting and helpful thought. When something goes awry, those who believe in God/karma/a loving universe have less of a reason to become stressed, because after all, a positive power is taking care of the larger picture. Indeed, a lot of research has linked higher religiosity with greater well-being.

But the people I encountered online seemed to see nothing but evil everywhere they looked, and they believed in it with a religious fervor. The plots designed to perpetrate this evil are never simple. They usually involve someone who wants to control or destroy something using a technology that appears benevolent but is in fact the opposite, in a very complex and intricate way. Interestingly, you don't find large numbers of people on encrypted social media platforms talking about the secret plot to improve the world, fight poverty and illness, and save millions of lives. No, the plots people focus on always involve things like taking away human rights, child abuse, corruption, controlling actions and thoughts, and the enrichment of villains.

I've thought a lot about these questions, and I've come to the conclusion that finding a villain or villains is not people's actual intention, nor do they set out looking for evil plots. Rather, I believe that chasing villains is a regrettable side effect of the way our psychological system deals with stress and fear. What I've come to understand about the complex emotional and psychological needs that drive people into misbelief is that their needs are initially, partially, and temporarily met by fixating on villainous people who operate in a morally black-and-white universe. In their deprived state they seek relief, and they find it in the form of a villain that is to blame for everything. And though the story of the villain can be complex, it makes the world seem clearly defined in simple categories of good and evil. Finding that villain is like scratching a mosquito bite; it provides only temporary relief. And in the long run it makes things worse—much worse.

In order to explain my thinking about all this, let's use another psychological coping mechanism as a metaphor: obsessive-compulsive disorder (OCD).

Obsessive-Compulsive Disorder as a Metaphor for the Funnel of Misbelief

Imagine a young woman—let's call her Amy—who was bullied as a child for being overweight. As a result, she developed concerns about how she looks—concerns that did not leave her as she grew up, went to college, and became a working professional for a local food supplier. Over time, the negative thoughts became more frequent and more repetitive, until eventually they became obsessive. She can't stop thinking about how she looks, especially when leaving the house to meet people. In trying to deal with these obsessive thoughts, she has developed a habit of checking herself in the mirror before she leaves home, touching up her hair, and washing her hands to make sure that she is ready. These behaviors do not resolve her concerns about her appearance, but they give her a sense of control and to some degree help calm her. Over time, these behaviors have become more routine and more compulsive, until she starts depending on them to get going. This is a simplified general characterization of OCD.*

Now let's take the general structure of OCD—the presence of obsessive thoughts and the compulsive behaviors that help make them more bearable—and think about the ways in which it could be a useful metaphor for the comfort that misbelievers seek. What I have in mind here is the idea that just as Amy finds a degree of solace in hand washing, so do misbelievers find solace in their preoccupation with finding someone to blame for the bad things that happen in their world. Feeling that they can make sense of events gives them some sense of control, at least for a while.

Of course, it's not a perfect metaphor, because there are some important differences between hand washing and finding someone else to blame for what is wrong in the world. But even the differences can help us better understand the deep needs that misbelievers are trying to satisfy by finding a villain and why this effort is doomed to lead them down the funnel of misbelief.

Hand washing is the same every day, but searching online for a villain changes and expands every day. Each time we look, there are new videos, more information, new connections, and more findings, which creates an ever-escalating trap.

Let's compare Amy to a young man named Tom. Tom isn't stressed and worried about his weight, he's stressed and worried about what is going on in the world around him. He is particularly worried about climate change. He knows people who have lost homes in wildfires and hurricanes. The stories scare him. The science confuses him. His thoughts become obsessive. He feels a desperate need to find answers for why all of this keeps happening. Naturally, he turns to the internet. It's important to note here that a behavior like hand washing wouldn't work for Tom. He is looking for an explanation, and because he is looking for an expla-

---

* In our competitive society, some people claim that they have "slight OCD," and sometimes it seems that they wear their self-diagnosis as a badge of honor, but the reality is that OCD is not a gift, it is a serious and potentially devastating condition.

nation for very bad things, an explanation involving a benevolent God will not cut it. No. Tom is looking for a bad mechanism—something evil that will explain and ideally take the blame for what is going on.

Tom finds a video that explains that climate change has nothing to do with burning fossil fuels; it's the result of secret geoengineering experiments being conducted by the government. He's shocked, but he's also calmer because he now feels he has a better understanding of the issue. The world makes a bit more sense, and that comforts him (much as Amy felt comforted the first time she washed her hands before leaving home). But this calmness doesn't fix Tom's stress and worry (much as in Amy's case), and soon he needs another dose of the same remedy (again, as in Amy's case). He goes back online, but this time his search points him in a slightly different direction, and he finds himself learning about something called the High-frequency Active Auroral Research Program (HAARP). It's a real program. He finds pictures on Google Earth of the former military research facility in Alaska. He learns that HAARP is capable of triggering hurricanes, earthquakes, and floods by vibrating metal particles in the atmosphere and that the story we hear about climate change is in fact a hoax designed to cover up HAARP's deadly impacts. It all begins to add up! Next he reads about how planes are spraying harmful chemicals and metals into the atmosphere that HAARP then "excites" with radio waves, altering the weather. He finds a post from someone whose brother is in the air force and confirms what he's read. Walking home, he suddenly notices how many jets are flying overhead, leaving plumes of white behind them. That night, he goes back to his research and learns that there's a reason for all of it: a plot to disrupt Earth's resources to such an extent that governments can use the weather as a weapon of mass destruction in a new cold war and seize greater control over their own populations as well.

In this sequence, you can see how Tom's coping mechanism—the compulsion to search for information online—is markedly different from Amy's coping mechanism of hand washing. Unlike hand washing, searching online for information is different every time; it sucks Tom deeper and deeper into the complex world of evil plots and villains and weaves an alternate reality. The deeper he goes, the more his outlook on everything changes. Tom is on his way down the funnel of misbelief.

Another important difference between OCD and misbelieving is that although the compulsion can become debilitating and harmful, hand washing is a neutral activity that does not create any long-term harm or change the person. Misbelieving can negatively change a person at their core, especially when the beliefs center around despicable characters who want to take over the world, destroy the rest of us, and control every aspect of our lives.

These two elements—the ever-expanding nature of the material available (particularly online) and the inevitable landing on an evil explanation—are two important features that differentiate the funnel of misbelief from OCD but similarly make the funnel of misbelief psychologically damaging and very hard to escape.

Obviously, finding an explanation that blames an imaginary villain is not a real solution to anything. It cannot help with a changing climate, a job loss, confusion, fear, anxiety, social neglect, and so on. It can, however, provide short-term relief. It creates a feeling of both knowledge and control, and in the short term this feels better. If you had asked Tom to rate his overall well-being level when he was in the midst of his confusion and concern, he might have said "Five out of ten." After he watched the first set of videos, he'd have said "Six." An improvement! Initially, when Tom finds a villain to blame, it is psychologically comforting. What a relief to feel that he understands what is going on and, more important, that it is not his fault. It is all the wrongdoing of someone else. But as he dwells on the ideas he's recently discovered and his worldview darkens, his well-being level drops below its starting point—let's say to 4 out of 10. So he asks himself: What could make me feel better? And he remembers what made him feel better the last time. He goes back online and finds more videos. Temporarily, he feels better again. But in the long term, he'll feel worse and worse. As he moves deeper down the rabbit hole of hate and obsessive thinking, the activity that gives him short-term relief changes his belief systems, slowly making him *more* concerned, and even obsessed, with the fears and anxieties he sought relief from in the first place (as illustrated in Figure 3). This pattern of short-term relief and long-term deterioration is part of what makes the funnel of misbelief so seductive and destructive.

FIGURE 3

The unintended consequences of finding a villain

Every time someone experiencing stress watches a video with a villain they can blame, they feel short-term relief that turns into long-term deterioration. The process repeats over time and creates a long-term reduction of well-being.

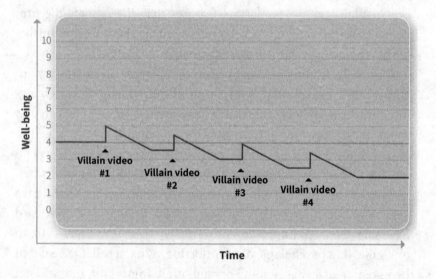

Intentional Hurt Hurts More

Let's take a closer look at what happens when someone like Tom finds a villain to blame and why it triggers a destructive cycle. As described above, Tom initially feels relief because he gains a sense of control; he knows why he's so stressed and anxious. It's not just random chaos,

HOPEFULLY HELPFUL

Learn to Enjoy the State of Ambiguity

Ambiguity is an uncomfortable feeling for most people. As such, it's one that we tend to quickly misattribute, concluding that something must be wrong. But in fact, ambiguity can be a productive, even positive state. When we conduct research, we try to encourage ourselves, our students, and our colleagues to enjoy a state of ambiguity. The logic is that when we don't know the answer, real knowledge can arise, so it is best to go slowly, think carefully, and enjoy the process of trying to find out what is going on. Sadly, this love of ambiguity is an uncommon mindset. Usually we want quick and clear answers, especially when we are stressed. For some people, this antipathy to ambiguity contributes to their descent into misbelief. The ability not to rush to conclusions, to keep multiple hypotheses in mind, and to remain open to new information and possibilities is key to not getting sucked into misbelief. We tend to admire and seek out conviction and confidence. But we would be better served if we learned to admire and enjoy a state of ambiguity.

The idea that we can learn to enjoy the state of ambiguity seems, at first glance, to go against human nature. After all, don't we want certainty? And doesn't any hesitation mean that something is wrong? Though a taste for ambiguity may not come naturally, it can be acquired, like other good things in life. Think for a moment about something you initially hated but developed a preference for over time. Maybe it was coffee, beer, experimental jazz, or bitter food. If we can acquire a taste for these things, why can't we develop a taste for ambiguity and learn to enjoy and embrace it? My sense is that this is not only possible but easier to achieve than a liking for hot sauce (which, by the way, directly stimulates our pain receptors but nevertheless we can learn to enjoy it). Once we acquire the taste for ambiguity, we may discover that it makes life more interesting.

it's because of an evil person. The corrosive feeling of being hard done by that we saw in Jenny and Eve's stories in the previous chapter has somebody behind it. In other words, *someone* is doing the deed. There's an intention behind the stress Tom is experiencing. This intention makes the relief he feels from finding a villain even shorter lived and eventually intensifies his stress. One beautiful study on the importance of intention and its relationship with stress was devised by Kurt Gray and Daniel Wegner. Their question was whether an aversive experience hurts more when we believe it's being inflicted on us intentionally.

Here's how the experiment worked. Imagine you're a college student and you've signed up to participate in this study. When you arrive, you're introduced to another participant—we'll call her Laura—and the two of you are told to sit in adjacent rooms. In your room, you see a computer screen that describes two potential tasks. One of the tasks involves what researchers call **discomfort assessment**, which means experiencing electrical shocks and rating how painful they are. The other does not involve any pain; it involves listening to two sounds and identifying which one is higher pitched. You're told that you will be completing one of the two tasks on the screen. However, you don't get to choose which of the tasks you will undertake. That choice will be made by Laura, who is sitting in the next room and can also see the same two tasks on her computer screen.

After a moment, you see the result of Laura's choice highlighted on the screen. She's decided that you're going to experience the electrical shock. You are somewhat annoyed, and you start to feel the fear of the looming pain. Why couldn't she have chosen the other task? What does she have against you anyway? You don't even know her. Annoyed, you start experiencing the shocks and rating your discomfort level.

Now imagine the same basic scenario, but with one key difference: when the researchers explain to you how the experiment works, they also tell you that, unbeknownst to Laura, the mapping of the buttons to the tasks has been reversed. In other words, you will receive the opposite task to the one Laura has selected for you.

When the selection is made, the electrical shock task is highlighted. You're annoyed, but you realize that it was not Laura's intention. In fact, she thought she was selecting the nonpainful option. The pain of the electrical shocks starts. The shock hurts. This time, too, you rate your discomfort level. But here's the question: Do you feel the intensity levels of the shocks differently in the two settings? In other words, would the same shock hurt less if you knew it was unintentional?

It turns out that the answer is yes. Intention matters, and the shocks feel different. Even though the physical intensity of the pain in the two scenarios was exactly the same, participants who thought their partner had deliberately subjected them to the shock rated the level of discomfort as higher than those who thought the pain had been inflicted without a bad intention. Intended pain hurts more than unintended pain. Suffering is a combination of external conditions and interpretation.

In addition, the experiment found that those receiving the unintended pain reported it as becoming less painful over the course of repeated experiences. In other words, they became habituated to it, as we tend to do with recurring experiences of all sorts. And the intended pain? Its intensity remained unchanged throughout the experiment. This selective adaptation (adapting to the unintended pain but not to the intended one) suggests that although we can become accustomed to many things in life, the malicious intent behind intentional pain makes it impossible to set it aside and give it less attention. It just continues stinging at the same level.

Hopefully, no one is sitting in the next room malevolently subjecting us to electrical shocks. But it is easy to find examples in our own life where intention adds an extra sting to a painful experience. I believe that happened to a lot of people during the Covid-19 pandemic. Like Tom, they felt stressed and anxious, went online searching for answers, and their discoveries convinced them that nefarious people were behind the virus and accompanying regulations. They felt temporary relief, but when the relief faded, they felt worse. It's bad enough to endure ongoing, unpredictable stress; it's much worse to endure it because of someone's evil intent. Unlike the participants in the intentional pain experiment, who had to endure the shocks only briefly, those of us suffering from the stress of Covid-19 related events (which means all of us to some degree or another) endured chronic stress for nearly three years. So when some people concluded that their suffering was intentional and as a result felt it even more intensely, what did they do? The same thing that made them feel better before: they searched for answers, for villains, for nefarious plots. And each time they found them, they moved deeper into the funnel.

Once a villain is designated, it is easy to return to them again and again to blame them for any new source of pain. When we're convinced that someone has evil intentions, we don't feel it's necessary to interrogate ourselves each time we pin a new crime on them. Quite the opposite; we point the finger unquestioningly. When Tom reads about another adverse weather event that scares him or comes across statistics about rising temperatures, it's easy for him to blame HAARP or the power-hungry government. And if you think you're different, think for a moment about a politician you strongly dislike. I'll use Donald Trump for this example, since a substantial portion of the US population would probably pick him, but you can replace his name with another if it better suits your political persuasion. Ask yourself: If I hear a story or read a headline that tells me Donald Trump is having an affair with a Russian prostitute or making secret deals with Putin, what's my first response? A lot of people, if they're honest, will admit that they believe the stories without independently verifying them. Why? Because many people have already decided that Trump is a villain. And all of us—wherever we fall on the political spectrum—love to have villains to blame.

HOPEFULLY HELPFUL

Don't Assume Malice

When bad things happen to us and we think that they are done intentionally, we suffer even more. Personally, I try to hold on to a slightly modified version of the principle known as Hanlon's razor. The original Hanlon's razor states, "Never attribute to malice that which is adequately explained by stupidity." Stupidity, however, is a harsh term, and I believe that the original intention of the saying was not really about stupidity but about irrationality and human fallibility. Here's my slightly modified version: "Never attribute to malice that which is adequately explained by human fallibility." This suggests that when bad things happen to us and the people we love, we should look farther and deeper into the underlying reasons for the events and seek explanations that don't involve intentionality or malice and instead involve mistakes, lack of consideration, impulsivity, emotions, or a whole range of other human quirks.

Believing in the slightly modified Hanlon's razor principle not only takes away some of the sting of intentionality; it also increases the odds that we will understand the real reasons for what went wrong and simplify the path toward improvement.

Why Do Misbelievers Gravitate Toward Such Complex Stories?

The stories spread by misbelievers are not just dark and negative; they are also surprisingly complex. Of course, in their minds, they are not stories at all. For example, it is not enough to claim that pharma companies want to increase their profits, so they took shortcuts with the development and testing of the Covid-19 vaccine or they are hiding information from the FDA and from the public (just to be clear, there is no evidence for any of that). Instead, the story is that Bill Gates and the Illuminati are conspiring to reduce Earth's population and create a technocratic society. To achieve this goal, they created the Covid-19 hoax and then a vaccine that will kill people with its side effects and eventually hurt both women's and men's fertility for generations to come. And the story doesn't end there. The fear created by the virus and the deaths will allow governments to quarantine their citizens and force them to carry a "green passport" that will convey their vaccination status—the first important steps into mindless obedience and the loss of individual rights. Of course, this is just the top-level plotline. It gets much more complex as we get into the details, such as claims about the way the mRNA vaccine works; allegations that Bill Gates owns Pfizer and has control of the World Health Organization; and various theories about the role of the central banks and digital currencies in controlling citizens of all nations. It is truly a never-ending story, and if it didn't have such a devastating impact, it would have been the best soap opera ever.

When I first encountered these narratives, I was baffled by their complexity. In general, my observation has been that humans have a strong preference for simplicity in the stories we use to explain the world around us. Just think about how many love stories follow the simple structure of a fairy tale with a happy ending or how many superhero stories have the same relatively predictable beats.

This preference for simplicity is also an important guiding principle in science, often packaged in the heuristic known as Occam's razor. The basic idea of Occam's razor is that the simplest explanation with the fewest moving parts is the one we should favor, until it is proven inadequate. Just to be clear, Occam's razor does not say that the simplest explanation is always the right one, only that when we don't have any data to help us pick one explanation over another, we should pick the simplest one.

Favoring simplicity has guided me in much of my own research. In my view, one of the goals of social science is to find a few central principles that can explain as much as possible about the world around us. For example, the principle of the **pain of paying**, which posits that it hurts to part with cash but it hurts less when we don't see or pay attention, helps us understand why we overspend when we use credit cards; why we feel worse at the end of a meal when we pay with cash compared with a credit card; why we sometimes prefer all-inclusive vacations even if they are more expensive; why we often go over budget when we renovate our homes; and much more.

With all that simplicity has going for it, why are misbelievers' stories so complex? Not only do they fail to adhere to Occam's razor, they actively complicate their theories with more moving parts and more indirect forces at work.

Take, for example, one of the most fascinating conspiracy theories: the belief that the earth is flat. Why do I find it so intriguing? Because for the earth to actually be flat, the magnitude of the conspiracy would have to be enormous. Every government would have to know the truth but hide it. Every pilot who's ever flown would have to see the truth of the earth below them and also hide it. The space program, of course, could not be real, and everyone involved with that operation would know the truth but keep it a secret. And of course, there would have to be lots of people working on the dome surrounding our flat Earth—moving the sun and the clouds, creating wind, and so on.

And that's not all. A subset of the people who believe the earth is flat have added another, more complex twist to the story: they are convinced that Australia doesn't exist. That's right: an entire country and the 26 million people who inhabit it are in fact part of an elaborate hoax. The Australians we see in the media are actually actors paid by NASA to pretend to be Australians. Or they're computer generated. And if you're thinking "Wait a minute, that's ridiculous, I've *been* to Australia," they'll tell you that you were in fact flown to a remote part of South America populated by the aforementioned actors. But *why*? Why would anyone go to such lengths to create a fake country? Well, the misbelievers have an answer for that question as well. Apparently, it all goes back to the British, who wanted to get rid of their criminals. You probably learned in school that all of those lawbreaking Brits were shipped Down Under, but according to the Australia skeptics, they were actually just dumped in the middle of the ocean to drown. The idea of an island on the other side of the world populated by bouncy kangaroos, cuddly koalas, and hardened criminals was created to cover up that act of mass murder. When I read about this theory, I wrote an email to my Australian friends, saying "I'm so sorry, our friendship was nice when I still thought you were real." Interestingly, the actors that play my Australian friends responded, insisting that they were real. Some people refuse to admit the truth even when they are found out.

The sheer number of people, governments, and institutions that would have to be involved makes the flat-earth conspiracy a fascinating one. Just think about what happens if you tell a secret to even five people. The odds that the secret would leak are very high. Or as Benjamin Franklin famously said, "Three may keep a secret, if two of them are dead." Yet for this conspiracy theory to be true would require hundreds of thousands or maybe millions of people keeping a secret.

When considering why conspiracy theories are so complex, one could argue that it's accidental—that conspiracy theories are unnecessarily complex because of the way the stories are constructed and the number of people involved in constructing them. But a more psychologically driven approach would try to understand the possible psychological benefits that come bundled with such complexity.

First, it is clear that a more complex story has more details and nuances, meaning that there is more to tell. As a consequence, new information can be revealed over time. So the people who benefit from the storytelling continue to benefit over a longer period. From this perspective, if you wanted to create a media empire around a misbelief (as the popular far-right radio host Alex Jones did, for example), you would want to make it complex enough that it can evolve over time. Just imagine a website that disagreed with the government's economic policies or positions on education. How often would people go to that website? At best, they might glance at it once, form an opinion, and move on. On the other hand, imagine a website exposing a secret cabal of pedophiles that included a changing roster of Hollywood power players and other members of the world's elite. Now you have a story that can be augmented, expanded, and updated. You can have new revelations about the members of the cabal and their actions, the underground networks by which they traffic their victims, the exotic private estates where they commit their crimes, and the elaborate cover-ups they use to protect one another. You can also offer details about the battle between those in the know and the rest of the world. People will keep clicking and clicking and clicking and sharing and sharing and sharing.

The second psychological reason favoring complexity is a cognitive bias called the **proportionality bias**. Proportionality bias is the idea that when we are faced with a large event, we implicitly assume that such an event must have been caused by proportionally large causes. The reality of life is that often "shit happens" without any rhyme or reason. Randomness and luck (including bad luck) are important forces in the universe, as is human stupidity, but this is an unnatural way for us to think. We look for reasons, for causes, and when something is larger, we look for larger causes. For example, imagine that tomorrow morning we all learn that putting fluoride in the water has made us want to see other people less frequently, isolate ourselves, and spend more time in front of our computers. What would be your first thought? That this was an understandable human error? Or would your mind immediately start thinking of everyone who might have gained from such an effect and the ways that they could have influenced our world? My guess is that you would at least consider the possibility that behind something as large as this there was a more deliberate evil intent. Consider the following examples (unlike the fluoride example, which I invented, this list includes topics that many people attribute to a large and malevolent cause).

For each pair of sentences, please check the box next to the one that you believe is more likely.

❑ HIV/AIDS was caused by a random mutation.
❑ HIV/AIDS was designed originally as a biological weapon and somehow escaped the lab.

❑ Hurricane Katrina was a freak event.
❑ Hurricane Katrina was a result of a weather manipulation technology that the government has been experimenting with.

❑ Mass shootings in the United States have all of a sudden increased in frequency.
❑ Mass shootings in the United States have been staged by the US government to advance a gun-control agenda.

❑ Princess Diana died in an unfortunate traffic accident.
❑ Some individuals with nefarious intentions had Princess Diana killed.

❑ TWA flight 800 was brought down by a simple fuel line malfunction.
❑ The downing of TWA flight 800 is a much more complex story involving the US military.

❑ The catastrophic California wildfires in 2018 were started by extreme fire conditions, faulty power lines, or lightning.
❑ The catastrophic California wildfires in 2018 were deliberately set by PG&E in conjunction with an international cabal of Jewish bankers led by the Rothschilds, using a space laser, in order to clear land for a high-speed rail project.

If you found yourself even entertaining or leaning toward the second answers in any of these cases, you may have exhibited proportionality bias. Here's one more example to consider: What about a virus originating in Wuhan, China, that led to a worldwide pandemic, causing almost all governments to announce quarantines, shutting down schools and travel, and creating tremendous emotional, mental, and financial fallout? If you learned about something like this, would you start wondering about the various governments or corporations that might stand to gain from such events? Or would you just chalk it up to bad luck?

Interestingly, the proportionality bias does not seem to apply to positive events. When amazing inventions are developed, such as penicillin, Post-it Notes, X-rays, Teflon, Viagra, and many others, we're very comfortable attributing them to chance. In other words, when it comes to major good things, compared with major bad things, we are much likelier to believe that "shit happens."

The third psychological reason favoring complexity is the desire for unique knowledge. As we explored in the previous chapter, feeling a loss of control under stressful conditions is a basic element in the funnel of misbelief. Individuals who have a low level of control over many important life outcomes have a deep need to regain some feeling of control. Importantly, one way to do so is to feel that they possess unique knowledge that sets them apart from other people. This psychological process works in two slightly different ways. First, the feeling of having unique knowledge alleviates some of the feelings of being powerless and at the mercy of hidden forces. Second, misbelievers are often not fully integrated into society. They often feel somewhat ostracized, as if society, including people close to them, looks down at them with disrespect and abhorrence. In such cases, the feeling of possessing unique knowledge provides a sense of control and feelings of superiority. In this way, unique knowledge creates a reversal, and the misbelievers feel that they are the enlightened ones with insight into what is really going on while others, the sheeple, are inferior. And maybe one day the sheeple will see the light and also gain access to that important knowledge.

In my many discussions with misbelievers, discussions that sometimes took hours and occurred over the course of several years, I often felt that they were spending so much time with me because they hoped they would be able to convert me to their worldviews. They believed that once I saw the world the way they saw it, it would certify their superiority and their special status.

Together, all of these forces pull the misbelievers away from Occam's razor and closer to Macco's razor (the complete opposite of

---

HOPEFULLY HELPFUL
Three Razors

To counteract the attraction toward complex stories featuring evil villains, it is useful to keep three "razors" in our cognitive hygiene bag. The term *razor* is used to describe certain heuristics, or cognitive shortcuts, that can help to quickly "shave" away unnecessary information and complexity and get us more quickly to the truth. We discussed two of them above: the slightly modified Hanlon's razor, "Never attribute to malice that which can be adequately explained by human fallibility," and Occam's razor, "The simplest explanation is the one we should favor, until it is proven to be inadequate." To these we can add Hitchens' razor, named after Christopher Hitchens, the late literary critic, journalist, contrarian, and staunch atheist: "What can be asserted without evidence can also be dismissed without evidence." Together, these three tools can prevent us from falling into a spiral of misbelief. They invite us to ask questions such as: Is it reasonable to assume malicious intent over stupidity, human fallibility, or chance? Is it sensible to propose a complex web of ill intention? Do I have the necessary evidence to support such an extraordinary claim? If the things we're trying to explain don't pass the test of these three razors, it's a sign that we should take a step back and suspect that we're onto the wrong explanation. We can also use these three razors in conversations with others to challenge their biases toward intentionality, complexity, and insufficient evidence.

Occam's razor), which states that "The most complex solution that involves the most devious intentions and the most hidden elements is almost always the truth."

The Psychology of Hate

As I began to understand the emotional and psychological forces that make us seek a villain and attract us to complex and nefarious stories, I also thought a lot about another powerful emotional element of misbelief: hate. When many misbelievers reach a certain point in the funnel, it's not enough to simply believe that there's a villain or an evil plot; they feel compelled to actively express hatred toward the alleged perpetrators and to inspire that same feeling in others. Hatred is not simply a more intense version of dislike; it's characterized by moral beliefs and it's associated with moral emotions such as contempt, disgust, and anger. I saw this in posts like the one that opened this chapter, in which Jon used graphic and shocking images to portray me as a mass murderer and was rewarded by a chorus of commenters who wanted to see me tried, tortured, or publicly hanged. I saw it directed at other people as well. As the comedian Bill Maher has pointed out, "The act of hating people you don't exactly agree with has become so ingrained, so routine . . . so normalized that now we don't even notice how often someone online is wishing someone dead. Anyone we disagree with about anything is evil incarnate, and every argument goes from zero to homicide."

What Maher is describing, and what we see around us, is not just hate—it is advanced hate. Hate 201. Hate 201 also goes by the name **moral outrage**, and is different from run-of-the-mill hate in a few ways. First, when a person expresses moral outrage, they elevate their own status from that of a hater to that of a righteous crusader for justice and morality. Second, by expressing moral outrage, the person might also be taking part in **virtue signaling**, whereby they signal to those around them that they are such highly moral individuals that they get outraged when other people don't meet their moral standards. Third, when a person expresses moral outrage, they eliminate any hope of forgiveness. Once someone is morally objectionable, there is no path to redemption. A person who has committed a crime can serve their sentence in jail and then turn a new page, but a person who has committed a moral crime is unlikely ever to be forgiven or allowed to start over. This way, moral outrage is not just directed against the particular action in question, it calls into question the whole person, now and forever. Fourth, framing something as a moral outrage calls other people to the flag and it increases the group's stamina in fighting the moral battle.

Understanding Hate 201 can also help us answer another question that once puzzled me: Why do the misbelievers' stories tend to feature not just selfish or greedy protagonists, but morally reprehensible crimes like pedophilia and satanic child sacrifice? Indeed, I've often wondered why these people are so obsessed with the idea that the elites are sexually abusing children, drinking babies' blood, and so on. I've also wondered why misbelievers like to call every villain a Nazi. But when I think about the nature of moral outrage, I see these accusations as designed to fuel the powerful moral emotions of disgust and revulsion, roiling this advanced form of hatred, which deepens the rift between the misbelievers and those who don't see the world the way they do. When we see someone expressing themselves with moral outrage, it's a sign that they've gone quite deep into the funnel. Their world has been divided into us versus them, with "us" righteously defending the moral high ground against the immoral, despicable, irredeemable "them."

At one point, early in my quest to understand this part of the story, I had a Zoom call with someone who declared that not only did he want me dead, he would personally like to be my executioner. His name was Richard. He was an ordinary-looking fortysomething guy who had been introduced to me by another misbeliever I'd met online. She'd told me that Richard had a podcast and wanted to have an "honest conversation" with me. She assured me that he'd ask difficult questions and be truthful and fair. I figured that maybe it would be a way to get my side of the story out there, so I agreed, which was how I found myself in one of the most unpleasant encounters I've ever had. Richard didn't ask me questions at all; he just shouted his version of reality at me. When I calmly attempted to correct him, he shouted even louder that my calmness was further evidence of my involvement in the conspiracies. He wouldn't let me talk, and when I tried, he got angrier. I got flustered, and worried that I would appear defensive. I tried to stay calm, but my emotions got the better of me. At one point, his eyes lit up as he caught some small discrepancy between something I said and something some reporter had erroneously written about me. "You're lying!" he exclaimed. "I saved this earlier, so I could catch you in the act!" At that moment, I realized that he had no interest in anything I might say; he only wanted to express his hatred of me, to look tough in front of his many followers and maybe pick up more followers and support in the process.

The conversation left me shaken and deeply disturbed. Yet there was something about the man's passion, however misguided, and his apparent intelligence that stuck with me. He was exactly the kind of person I wanted to understand: someone who, once upon a time, had started out with a relatively reasonable and truth-based worldview but had descended down the funnel of misbelief to a point where he spent all his time digging through dark and disturbing stories, sharing them with others, and yelling abuse at people like me. Once I decided to get deeper into this world, I reached out to Richard and requested another conversation—this one with me interviewing him—and he agreed. To this day we have kept in touch, mostly via Telegram. In many of our discussions, he stopped shouting and was quite forthcoming. Over time, I came to think of Richard as my personal "spirit guide" to the underworld of conspiracy theories and misbelievers. My relationship with him gives me a useful window into the psychology of hate and the perspective of the misbelievers.

HOPEFULLY HELPFUL

Don't Be Part of the Silent Majority

"If you don't have anything nice to say, don't say anything at all" is the kind of advice we used to get from our grandmothers. Unfortunately, the internet seems to work on the opposite principle: it's the people who have something nice to say who tend to be quiet, while those who are angry, hateful, and full of negative feelings are much more motivated to express them. When I was under attack by the misbelievers, terrible things were written on my personal Facebook wall every day. I know that they were seen by many, many people who knew that they were untrue. Some of those people reached out to me personally. But it was very rare for anyone to say something publicly—to post a comment or a reply that retaliated against the hate. Of course, on one level this is understandable, because the moment we speak up, we risk becoming a target of hate ourselves. But the sum total of all that silence has long-term negative consequences and makes the internet look like a much more hateful place than it really is. In reality, the haters are only a small percentage of any given population, but they take up a lot of space. So my lesson from this is that when we have something nice to say, we should say it.

Entertainment? Yes, Entertainment

It is hard to describe how painful I found the initial attacks. I'd never been a target of hate like that before. Now, more than two years after the attacks started, I've gotten used to them to some degree (yes, we can even get used to death threats); learned to live with them to some degree (don't look at anything from the misbelievers first thing in the morning and before going to bed); and even directed my energy to studying and understanding them (most likely as some semifunctional coping mechanism).

Throughout this time, I've spent many hours a day looking at their information from various sources and formats. Websites, videos, papers, posts, reactions to posts, tweets, Facebook Lives, videos on Rumble (an app for those who have been kicked off YouTube), pieces of information that look like they're from legitimate organizations, pieces of information that look like authentic news shows, pieces of information that look like academic papers, pieces of information that look like documentaries, and more. In the process, I've become used to spending time with this information and even become familiar with the voices of a few of the individuals who spread a lot of the misinformation. Why am I telling you this? Because I recently realized how engaging I find their material.

How did I come to this conclusion? I decided to take a few weeks to work on this book alone in the mountains. I rented a small cabin with a desk, and every day I did nothing but read academic papers that are related to this topic, dictate my ideas (usually while walking), and immerse myself in even more misinformation than usual. With all my time dedicated to this topic, my only choices were between reading academic papers, writing, and engaging with their materials. Nothing else—no deadlines, no emails, no conference or video calls. And guess what? Much like an addict, I felt more and more attracted to the disinformation. This is not to say that I believe much of it (though from time to time they do voice a truth, even important truths). Rather, my point is that the delivery and the nature of the storytelling are so captivating that it is hard to stop consuming the material.

With that in mind, I thought back to the video that opened with me in the hospital and went on to explain how my injury made me hate healthy people and want to reduce the world's population. I now realized how engaging it must have been for other people. In fact, I realized that I myself had enjoyed all kinds of material that must have been equally abusive and offensive to the people described therein.

There are three points that I want to draw from all of this. First, evil people, evil intentions, and engaging plots provide wonderful starting points for interesting stories. Facts are rarely as entertaining as fictions. Just watch Oliver Stone's movie *JFK*—it's a compelling drama that makes us *want* to believe the theories it dramatizes, despite the questionable historical evidence for most of them. Second, the people who produce the information for the misbelieving crowd know what they are doing in terms of storytelling. And not having to remain truthful gives them an additional edge in terms of the potential plot outcomes. They can break news every day. They don't have to wait for something to actually happen. How can science—which is slow and methodical, providing only an occasional breakthrough—compete with creative minds unfettered by facts? And third, the competition for human attention between accurate and true information on the one hand and false, sensational information on the other is not a fair fight. As we've seen in this chapter, the proportionality bias drives us toward elaborate stories. The stress we feel cries out for a villain to blame and for clear moral lines rather than the messy shades of gray that often represent reality. Take anyone with time to spare, and they are likely to watch way too much "burned-person-hates-the-world-and-wants-to-destroy-it" content. Even in a beautiful, peaceful cabin in the mountains.

Speaking of entertainment, part of the challenge of misinformation is that it is less like a book or a movie that we passively consume and more like a game in which we participate. Indeed, this is one of its most seductive and overlooked qualities: misbelief is enormously engaging and even fun for those who become deeply involved in its cleverly constructed alternate worlds. People who work in the gaming industry have drawn striking parallels between the structure of a conspiracy theory such as QAnon and the structure of popular alternate-reality multiplayer games. QAnon is "addictively participatory," wrote Clive Thompson in *Wired*, with its oblique clues, hidden messages, and secret platforms. ". . . To belong to the QAnon pack is to be part of a massive crowdsourcing project that sees itself cracking a mystery." It's thrilling, energizing, and creative. People don't just absorb the information, they add to it and generate their own content. They feel important, because they can play a role, through their own dedication and intelligence, in outsmarting the evil elites who are controlling the world. The game designer Reed Berkowitz has called QAnon "the gamification of propaganda." It is, he wrote, "a game that played people."

HOPEFULLY HELPFUL

Be Suspicious

One of the tricks of fake news or misinformation is that it is often designed specifically to feed on human weakness and to get us excited about the story unfolding in front of us. This has a few important implications. First, when we consume information, we need to put on heavy-duty suspicion glasses. This is not a fun way to live, I realize, but it is very important. It is particularly important to be cautious before we share information and unintentionally contribute to the further spread of fake news. Second, before sharing anything, we need to understand that we're not just passing the information along, we are attaching our reputation to it, giving it our stamp of credibility, as if we vetted and are recommending it. Would you enthusiastically recommend that a friend purchase something you'd never tested yourself? Probably not. But you may be unwittingly doing this with information every day. Again, suspicion and caution aren't the most pleasant lenses through which to view the world, but in the realm of information we would all be served well to adopt them more often.

Unfortunately, however, when it comes to misbelief, the line between the real and the virtual is dangerously blurred. If misbelievers were just entertaining themselves in private chat rooms, perhaps their activity would be no more harmful than people playing Dungeons & Dragons or Grand Theft Auto. But this is where the parallel breaks down. It's all fun and games until someone starts shooting Black people to prevent them from outnumbering white people in a "great replacement"; or someone harasses the grieving parents of mass shooting victims because they think mass shootings are a hoax to take away guns; or people storm the US Capitol in an effort to subvert an election they believe was stolen. Even those who aren't committing public acts of violence may be taking small actions in the real world with real people, with consequences that are anything but entertaining.

In Summary . . .

In this part, we've examined the emotional element of misbelief—an appropriate starting point, since emotions are often the underlying driver for our beliefs. As we've seen, the impacts of stress are deep and far reaching. But keep in mind that these elements are not sequential stages. Emotions play a role at every step in a person's journey down the funnel. Next we'll turn to some of the cognitive elements, examining the ways in which our rational capacities can be used in highly irrational ways to form and confirm beliefs. Then we'll take a look at the underlying personality traits that might make some of us more predisposed than others to becoming misbelievers. Finally, we'll consider the social factors that draw people into communities of misbelief and keep them there, reinforcing and even escalating their convictions.

FIGURE 4

The emotional elements of the funnel of misbelief

·· We all experience general
stress—predictable and unpredictable.

• The experience of unpredictable stress
can create a feeling of lack of control,
which can lead to the condition of
learned helplessness.

·· **Stress negatively impacts cognitive
function and decision-making.**

• Resilience is increased by community
support but weakened by economic
inequality.

·· **Under compounding stress, we can
begin to feel hard done by.**

• This leads us to seek answers and ways
to regain a sense of control.

• We get a temporary sense of relief and
control when we find a villain to
blame—similar to the relief an OCD
sufferer gets from compulsive behaviors
like handwashing.

• However, the relief is temporary, so we
go back for more. In the long term, it
makes us feel worse, but we keep looking
in the same place for short-term relief.

·· **Feeling as though the pain is being
inflicted intentionally adds to the sting.**

·· **The stories misbelievers believe are
extremely complex for several reasons:**

• For the content creators, complex stories
mean they can produce more content.

• Complex stories satisfy the
"proportionality bias," which tells us that
a large or intense problem must have
large causes.

• Complex stories satisfy the desire for
unique knowledge, making the
misbeliever feel empowered and more in
control.

·· **The stories often have a morally
repulsive theme, designed to fuel hate.**

PART III

THE COGNITIVE ELEMENTS AND THE STORY OF OUR DYSFUNCTIONAL INFORMATION-PROCESSING MACHINERY

CHAPTER 5

Our Search for the Truth We Want to Believe In

It ain't what you don't know that gets you into trouble. It's
what you know for sure that just ain't so.

—MARK TWAIN[*]

How are misbeliefs created? How do we become convinced that particular pieces of information are, in fact, facts? How do we take those pieces of information and add them up into a cohesive narrative? And how do we become so sure of our narratives of choice that we allow them to shape major decisions about things like our health or our politics? How do they become so important to us that sometimes we are prepared to fight with family members and lose friendships over who believes what? We would never end a relationship with someone just because they prefer the color green to the color orange (I know, can you believe people?), but beliefs about stolen elections or the curvature of the earth can drive a deep wedge into our relationships. And why is it that when we adopt such narratives, we feel so righteous that we are compelled to share our beliefs and spread them?

---

[*] Or perhaps not. You can't trust everything you find on the internet. This quote, though widely attributed to Twain, can't be confirmed to be his. The interesting question is whether, a month from now, you will remember this quote as if it were Twain's.

In the earlier chapters, we examined the emotional experiences that set some of us up for misbelief. We looked at the way stress makes us feel out of control and hard done by and why some people seek a villain to blame for their suffering. And we've seen how this coping strategy provides short-term relief but, over the long term, pulls people into the funnel of misbelief. It is at this point that the cognitive elements take center stage. When stress drives a person to seek relief in the form of answers, a series of cognitive processes guides that search and shapes the way that the information (or misinformation) is processed. Before we jump in, a warning: if we want to understand how a misbeliever thinks, we need to be prepared to confront some uncomfortable truths about the objectivity of the way we all think and process information. In this chapter, we'll examine several common facets of human cognition that impact the way we form and maintain our beliefs. But first let's take a look at the general problem of misinformation and how it interacts with the mind of the misbeliever.

When Misbelief Meets Misinformation

In this day and age, it's impossible to talk about misbelief without also talking about misinformation and the channels that distribute it; in particular, social media. Much has been said and written about the ways in which social media's algorithms are designed to take advantage of the peculiar weaknesses of the human mind, as in an unhappy but deeply codependent marriage. Behind every post spreading misinformation there is a range of interests, as discussed in chapter 1, and each of these agendas, whether intentionally or unintentionally, plays a part in driving the spread of misbelief. Add to this advanced AI tools that can generate ever-more-convincing misinformation in greater quantities, and it seems like we'll never get a handle on the problem. But it's important to remember that misinformation would not be nearly so effective were the human mind not so susceptible to these forces.

In fact, social media would never have become so popular without being built to take advantage of the faulty circuits affecting the way we react to information.

My primary focus in this book is on the human side of the story, on understanding what makes us susceptible to this kind of information. I will not spend time discussing the well-documented ways that social media technology is designed to take advantage of us. But social media and the faulty circuitry of the human mind do go hand in hand, and it's no accident that misbelief has flourished in the age of social media.

Consider the following analogy: Humans have evolved to crave fat, salt, and sugar. This was all well and good when the available sources of satisfaction were to be found in nature's bounty—fat from nuts, seeds, or occasional meat and fish; salt in certain plants or seaweeds; sugar in fruit. Our cravings used to serve us, helping to ensure that we consumed the range of nutrients we needed to survive. But then came the age of modern food processing, and the production system began to create products that perfectly appeal to our evolutionary predilections. *The doughnut! The French fry! The burger!* The fast-food industry's success has relied on hijacking our innate cravings to persuade us to eat substances that are not nourishing, healthy, or satisfying in the long term but maximize profits for the companies that sell them in the short term. If you're inclined to believe in conspiracy theories, you might start speculating that in reality it is the pharmaceutical companies, which make diabetes medications and statins, that are plotting our feeding regimens. Or if you have been reading carefully, maybe you suspect that it is Bill Gates, the Illuminati, and me who came up with this dietary disaster, as part of our effort to reduce the world population.

What is obvious in the story of modern food is that if we want to stay healthy, avoid chronic disease, and live a long life, we must learn how to negotiate the relationship between our appetites and the plethora of available options. The same applies to our relationship with information. The human mind has evolved to quickly navigate the complexity of our world; create and sustain tribal bonds; establish beliefs; make rapid decisions; and do this in an imperfect way that also comes with the cost of a range of biases, shortcuts, blind spots, and other quirks. That was all more or less fine for a while. But the internet and social media have come along like the information equivalent of highly processed fast food, hijacking those cognitive quirks and tempting us into beliefs and behaviors that are perhaps even less healthy for our minds than doughnuts are for our bodies.

Anyone who interacts with social media falls prey to this to some degree, just as we almost all eat at least some French fries when they are brought to our table. (Okay, maybe you don't. But I certainly do.) But is there more we can do to protect ourselves from the temptation?

Andy Norman has made the analogy between misinformation and a virus and has discussed the importance of "inoculation" as a way to defend ourselves against infection by misinformation. The analogy, of course, is not perfect, but it is important to consider the ways in which we can protect ourselves and our kids from infection. As Norman suggests, developing higher levels of media literacy and a deeper understanding of the mechanisms that misinformation creators utilize (activating emotions, using outrageous terms, employing anecdotes) can help us protect ourselves. Learning to recognize when a story is anecdotal or when it is designed to trigger an emotional response, for example, can improve your chances of not getting sucked into it.

Inoculating yourself against misinformation is not a one-shot deal, unfortunately; it takes practice. In my lab, we built and tested disinformation games designed specifically to help people learn to spot the tricks of the trade—that is, the tricks of those who engage in spreading misinformation. In those games, various characters shared real and invented news. The participants were tasked with differentiating between the two types and the strategies used to make the misinformation more appealing and more tempting to share. For example, a character we named Ann McDotal shared inaccurate information that relied on emotional storytelling. Here's one of her posts.

OMG everyone, my friend just told me that her cousin's husband's niece was diagnosed with HORRIFIC AUTISM after she got vaccinated against measles!! Can you believe it? SHARE IF YOU ARE TERRIFIED!

Another character, Dr. Forge, propagated fake news by boosting his false expertise and pseudoscience. His posts always referenced a person with multiple letters after their name, like this:

Like and share please: Dr. Singh (PhD, MD) admits that research on toxic chemicals in common vaccines is being swept under the rug by Big Science!

Mystic Mac simply never met a conspiracy theory he didn't like. His posts often linked together multiple theories, like this:

Psst . . . This is serious: a cabal of Big Science and government high-ups have colluded for years to promote global vaccination programs to gain power. Look into it. The government is 100% in on it.

And Ali Natural relied on the naturalistic fallacy—the notion that if something is natural, it must be good. She'd post claims such as:

Don't let doctors force vaccinations on you or your children! Physicians should be helping you boost your body's natural healing abilities! You can protect your immune system naturally, without chemicals, by eating pomegranate!

The games worked! After being exposed to those characters and their strategies during the game, the participants became better at identifying manipulative information and were also less likely to share such information on social media down the road.

We all can—and should—work to educate ourselves about the nature of misinformation and the many forms it takes. But it's not enough to focus on the problem "out there" and the false and mis-

leading content we encounter every day. We also need to understand our own predilections toward that mentally unhealthy menu
and learn to identify some of the cognitive biases, structures, hab-
its, and data-processing algorithms that make all of us susceptible
to informational junk food. As a starting point, let's take a closer
look at an example of a particular conspiracy theory and the ways
in which it interacts with the minds of the people who encounter it.

Anatomy of a Conspiracy Theory

There are, sadly, a multitude of examples I could pick to examine
the inner workings of conspiracy theories, from the reptiles walking
among us to the JFK assassination to many others. There may be
many things missing from our world, but there is no shortage of
conspiracy theories. Some are large and sprawling, playing out on a
global scale over decades and weaving together disparate people and
events, both current and historical. Others are limited to a specific
instance or place. The one we will focus on here is a relatively small
and compact conspiracy theory that struck me as a good case study:
the belief that the Covid-19 vaccines contain tiny magnets.

I first encountered it in a text from someone I know: "Can you
believe this? We have to do something about this. My friends are
all sending me videos of themselves magnetized." That alarmed
message appeared in my Telegram app and came from none other
than Maya, one of my misbelief spirit guides. I first met her during
a very unpleasant podcast interview with another misbeliever who
just wanted to shout at me. She later called to apologize, and we had
numerous less confrontational conversations. At some point I had
agreed to go onto her podcast, but the moment she posted about the
interview, her Telegram channel was deluged with so much hate toward her for even talking with me that she canceled the interview.
After that, we stayed in touch and she sent me various links to things
she believed were happening. This particular message linked to a

video that indeed showed a person with a piece of metal sticking to her shoulder at the exact same place where, she said, she had been vaccinated. There was also a link attached to another video that expanded on the alarming story.

By that time, Maya and I were old acquaintances and had even met in person. I followed her channel on Telegram, so I had seen the things she shared with her growing number of followers and knew that she was a deeply committed misbeliever. It appeared that she was becoming more dedicated to her worldview as time passed and was methodically advancing into the role of an influencer within the universe of misbelievers. Eventually she crowdfunded among her followers to support her "research," and a few short weeks later she became a full-time dedicated social media conspiracy theory researcher and spreader. Knowing all that, I was obviously disinclined to take her claims seriously (an example, perhaps, of my own cognitive biases, but there you have it; sometimes biases can save us a lot of trouble). However, I was curious as to why she was asking for my opinion about the magnet video and why she was so alarmed about it. I have to admit that I was also curious about what was occurring in the video and wanted the explanation for it. And so I clicked on the link and began my acquaintance with the magnet theory. Let me walk you through it.

At first glance, the video looks like a clip from any news network. Headlines and logos run across the bottom of the screen. A graphic of a city skyline fills the background. The host, Stew Peters, wears a suit and tie and speaks confidently into the camera. All of this gives the impression of a professional, well-financed, credible production—one that can be trusted. At this point, I pause the video and look up Mr. Peters. He is a former rapper and bounty hunter who has become a popular figure in right-wing online media.

After reminding viewers about the previous week's segment, featuring a brave and tragic "vaccine injury victim," Peters introduces the topic *du jour*. Social media, he declares, has been flooded with countless videos of people sticking metal objects to themselves. This reference to a "flood" of cases is not accidental. It activates what so-cial scientists call **social proof**: if lots of people call in to the show to back up a claim with their own stories, it increases the odds that the claim will be taken as true. "I've received so many emails about this magnetism since that interview, I can't even keep up," Peters claims. The people writing the emails are commended for their strength and bravery in speaking out, their "concern for the rest of us, particularly our kids."

Next there is video evidence of some of these heroes—homemade clips of people with metal objects apparently sticking to their bod-ies, similar to the video of Maya's friend. We see coins, keys, kitchen utensils, and more.

At this point, I stop the video and search online for more "evi-dence," and I find many videos, including a personal favorite that shows a baby with an iPhone stuck to his chubby thigh. And no, this is not one of the newer iPhones with the built-in magnet on the back! It is a standard old iPhone. Some of the videos show metal objects sticking to the body; some show magnets sticking to the body (wait, I thought the magnet was inside?); and some show coins and silver-ware (even though most coins cannot be magnetized).

Okay, that's far enough down the YouTube rabbit hole for me. Back to the Peters video. By now viewers have visual evidence to add to the social proof. Peters acknowledges that at first it seemed like a fringe phenomenon and medical experts were reluctant to connect it to the vaccine. But—no more! He has an expert guest who will tell all.

There's the magic word: *expert*. His guest, Dr. Jane Ruby, is intro-duced as an international health economist with over twenty years' experience in pharmaceutical research. Again, I stop to consult Goo-gle and cannot find any evidence of her having actual medical cre-dentials.

Dr. Ruby starts by describing a medical technology called *magne-tofection*, which uses magnetic fields. She invites the audience to look it up on PubMed, Wikipedia, and the government's websites. This is an important step. By inviting viewers to fact-check her claims, she adds to her perceived trustworthiness. After all, would a person who was not sharing actual facts do that? Surely someone who was trying to spread falsehoods would not invite people to check their statements!

At this point, you might decide to do as she suggests and check. Or you might not. Would the expert's encouragement to look for yourself make you more or less likely to do so? A few years ago I studied this question in the domain of medical advice. I examined how patients reacted to medical advice that was accompanied by a suggestion from their doctor to get a second opinion. It turned out that when the doctor suggested that they get a second opinion, more patients trusted the doctor more and with that increased trust, even fewer of them ended up getting a second opinion. My guess is that Dr. Jane Ruby's invitation would elicit a similar response. By inviting viewers to fact-check her, she ensures that fewer of them will and many more will trust her.

What about the viewers who decide to check? Well, I do. "Magnetofection" is an odd-sounding technology, and I'm curious and somewhat suspicious about it. But with a little help from Google, I learn that there is indeed something called magnetofection. It seems to be a promising technology in the field of gene therapy. Who knew?

Clearly Dr. Ruby knew. By choosing something obscure— something that viewers likely didn't know about—that turned out to be real, she further buttresses her credibility. Even I, with my high level of mistrust, am warming up to her at this stage. Perhaps intuitively, I conclude that if she knew about this technology when I did not, she must know other things I also do not. Maybe she is a real expert, someone on the frontier of medical science?

The choice of magnetofection as a starting point has another advantage besides being real: it sounds like something technological and unnatural, and it's easy to imagine that it could have a large potential for harm if it were to fall into the wrong hands.

Trust is now established. Those who have made it this far are most likely convinced of Dr. Ruby's expertise and knowledge and from here on are unlikely to check everything that she says. That's why it was so important that the first piece of information she shared was true. Now truth begins to give way to falsehood, fact slides toward fiction. But who's checking? Most likely, not the viewers.

The magnetic material, Dr. Ruby says, is available to purchase, but—"And here's the frightening part," she adds dramatically—it's not intended to be used with humans. (I keep fact-checking and learn that this is not true; many clinical trials of magnetofection have been carried out with humans.) Dr. Ruby claims that this magnetic material is part of the Covid-19 vaccine—acting as a "forced gene delivery system" that carries the vaccine into every corner of your body, including "places in the body that God and nature never wanted any foreign material to get into." (In case you were wondering, this is also not true.)

At this point in the interview, Dr. Ruby unleashes a flood of medical terms and acronyms that further add to her aura of expertise, accompanied by a medley of background graphics including spinning DNA, magnets, molecules, brains, vials of vaccine, and needles.

Peters now interjects to "address the fact-checkers," reading out a denial of the magnet claims from the website of the Centers for Disease Control and Prevention (CDC). This serves two purposes. His apparent willingness to consider a different view adds to his impression of trustworthiness. It also sets up his guest to give voice to what many of his viewers already believe about the CDC.

"Are they just lying?" he asks her.

"Yes, they're lying," she replies without hesitation, looking unblinkingly into the camera. "And Stew, you know why? Because the past is prologue." This somewhat enigmatic statement sets her up to reel off a list of the CDC's supposed past infractions, of which many of her viewers are likely already convinced. If those are true, this new finding must be true as well.

At the end of the video, Peters praises his guest's diligence and invites viewers to reach out to her at a secure, encrypted email address, enrolling them as fellow truth seekers and fact finders. And there you have it: a very effective twelve-minute video spreading a completely false narrative, delivered within a somewhat masterful construct that is designed to pull people in, win their trust, and persuade them to believe.

I like the magnet theory because it is a prototypical conspiracy theory. It starts by claiming evidence from lots of people. Trust is created. Next we are shown a range of videos demonstrating the phenomenon firsthand. Trust increases. How can anyone argue with video evidence? Then it introduces an expert with official-sounding letters following her name who builds further trust by saying a few correct things and reiterating many things that the audience already believes, then uses that trust to introduce one more (false) element into the viewers' belief systems. Finally, viewers are asked to take part in the quest for "truth." *Et voilà*, you have a beautifully packaged theory: a grain of truth, evil villains, a reputable expert, social proof, firsthand videos, and a request to be on the lookout.

Mind over Magnets?

You may wonder, what kind of person falls for this? Surely you wouldn't. You'd question the expert's credentials, look up her claims (even the ones she didn't invite you to check), and watch those homemade videos with a skeptical eye, right? You'd remember how the laws of physics work and conclude that even if there were magnets in the vaccine (which there are not), they'd never be powerful enough to hold an iPhone. (Have you not tried and failed to stick a takeout menu to the fridge with a small magnet?) You'd also think twice about the claims regarding Covid-19 vaccines. Even powerful pharma companies cannot reverse the laws of nature. And if you know anything about how heavily regulated businesses work, you'd realize that although pharma companies might not be the world's most beneficent corporations, there's no way that multibillion-dollar companies would risk their reputations by violating FDA rules. I suspect that like many other US companies, they have compliance departments almost as large as, if not larger than, their research and science departments.

It is easy to discount the belief-creating process on *The Stew Peters Show* as something only other people would fall for. But this is a good opportunity to examine our own beliefs and the process by which we came to have them. In reality, are we really that different from the people watching and sharing videos like the magnet-in-the-vaccine theory? Here's a simple exercise to find out.

For each of the statements below, please circle True or False according to your belief. Then rate your confidence level in that belief on a scale of 0 percent (not sure at all) to 100 percent (perfectly sure).

And yes, I know that at this point you are probably thinking to yourself that it is too much work to put down the book, grab a pen, and then come back to the book. You likely think that you will just go over the list of questions and remember your answers, but trust me, you will not. When we look at a question and don't commit to a specific answer, we have the impression that we have answered the question, when in fact we didn't. Later on, if someone tells us the answer, we can very easily convince ourselves that we got it right. But maybe we didn't. We just had the feeling that we answered the question. Avoiding this kind of mental hindsight response is the exact purpose of this exercise. So please go ahead, find a pen, circle whether you think each statement is true or false, and indicate your confidence on a scale from 0 (not confident at all) to 100 (perfectly confident).

Covid-19 is real: True/False. Confidence (0–100): _____
5G is safe: True/False. Confidence (0–100): _____
The universe started with the Big Bang: True/False. Confidence (0–100): _____
The earth is round: True/False. Confidence (0–100): _____
The Holocaust really happened: True/False. Confidence (0–100): _____
God exists: True/False. Confidence (0–100): _____

Now go back to each of the statements and ask yourself each of the following questions, one at a time, for each statement:

How did I come to believe my answer to the question about:

Whether Covid-19 is real or not: _____

Whether 5G is safe or not: _____

Whether the universe started with the Big Bang: _____

Whether the earth is round: _____

Whether the Holocaust really happened: _____

Whether God exists: _____

How can I be sure about my answer to the question about:

Whether Covid-19 is real or not: _____

Whether 5G is safe or not: _____

Whether the universe started with the Big Bang: _____

Whether the earth is round: _____

Whether the Holocaust really happened: _____

Whether God exists: _____

If you are like most people, you probably rated your initial degree of confidence in your beliefs fairly highly. But when you answered the follow-up questions, which required you to reflect on the process you used to reach your beliefs and your confidence in them, maybe you realized that your beliefs are not built on such solid ground as you initially thought. Maybe you realized that your beliefs are largely an outcome of accepting things as they were given to you. In reality, perhaps you haven't done much independent research or even much soul-searching about your beliefs. No doubt you've relied on established sources and experts you trusted. You've accepted social proof (the "everyone else thinks so" approach) or the firsthand testimonies of individuals (the "my friend told me" approach). None of this necessarily means that you were wrong, but recognizing the shaky ground on which some of our beliefs are built might reduce our confidence in our beliefs, just a little. But don't worry. The odds are that in thirty minutes or so, you will forget about this short period of self-doubt and return to your default state of high confidence.

My point is not to call into question all of our beliefs about historical events or global health crises. Nor is it to criticize us for not researching everything more diligently. I'm simply trying to highlight the fact that we all use shortcuts to deal with the complexity of the information-saturated world around us. It is just not humanly possible to doubt everything all the time. So we work very hard to set some basic parameters. We have a set of things that we believe and others that we don't. We choose experts we trust and find sources we think are credible. We can't revisit all of our opinions all the time. We need these basic beliefs to protect our sanity from the storm of data we encounter every day. Our basic beliefs act as a defense mechanism, protecting us from wandering around the world feeling that we have no idea what is going on. As more information becomes available online, we need more defense mechanisms, and as a consequence, we tend to become stuck in our opinions.

It's not just big beliefs like the origin of the universe or the existence of God that we arrive at in this way. Every day, our behaviors and choices are guided by beliefs we've adopted but perhaps not deeply researched or questioned. Take me, for example. A few years ago, I decided that I like the logic behind intermittent fasting. At that point, there was not much evidence for it, but it made intuitive sense to me and some friends told me it worked for them, so I started following it. Later, when studies came out showing that it had no advantage over other approaches to reducing calories, I always managed to find fault with those studies, and to a large degree I stuck to my original opinion (although I used the evidence from those other studies to justify my occasional late night out eating and drinking with friends: "Why keep to the strict rules when the evidence is just not there?"). Even as I write these words, years after I started following the diet, I am not sure if it is a good one or not, but I try hard to stick to it most days.

HOPEFULLY HELPFUL
Challenge Your Beliefs

Interrogating our own convictions is easier said than done, but a tried-and-true method was proposed by the ancient Greek philosopher Socrates. The Socratic method, as it is known, is based on posing a hypothesis and then challenging it. There are two ways to challenge a hypothesis: either by seeking alternative explanations or by testing whether it is falsifiable. For example, you might come up with a hypothesis that the 2020 US presidential election was rigged by foreign actors. But then you would ask yourself: What other explanations might there be for the result of the 2020 elections? Maybe a majority of voters simply preferred the candidate who won. And you might ask: What would it take for my hypothesis to be untrue? Perhaps the numerous recounts and voting safety measures in place would help undermine the hypothesis. In this way, the Socratic method can be used to interrogate ourselves and keep us honest. I'd recommend that you do it on paper rather than just thinking it through in your head. As discussed earlier, it's too easy to fool ourselves into thinking that we've answered a question in our heads when in fact we have not. The simple act of writing something down with pen and paper makes a big difference. The Socratic method can also be used in a conversation with a misbeliever. Rather than simply refuting or denying their hypotheses, it is better to question them about possible alternative hypotheses and falsifiability.

Another useful method for honestly questioning our own beliefs is to ask ourselves whether a neutral third party would come to the same conclusion as us. What if that person did not share our current biases in viewing the world or even perhaps held opposite views on the topic at hand? What would they say? How might they challenge our hypothesis?

Beliefs are powerful because once we have them, it takes a lot for us to question them. We're more likely to avoid the hard work of questioning ourselves and instead double down, just as I do with my diet. As the documentary filmmaker Adam Curtis has said, "All power in the world works not just through force or by laws—but also by getting inside people's heads and shaping how they see the world. This is true even in the modern age of individualism." I would go further and say that it is *especially* true in the modern age of individualism and digital information flow.

Biased Search: What My Would-Be Executioner Taught Me About the Misbeliever's Mind

Remember Richard, the guy who told me that he'd like to be my executioner? Clearly, our relationship didn't get off to the best start, but nevertheless, we stayed in touch. His obvious intelligence and his singular dedication to his "research" made him a fascinating travel guide in my quest to get inside the heads of misbelievers.

Richard presents himself as a truth seeker who is on a mission to expose the webs of falsehood and deception perpetrated by powerful elites, and I think that he deeply believes this. Covid denial was by no means his first conspiracy theory. Indeed, Richard has been a subscriber to many alternate realities, going back many years. He believes that Bill Gates controls the World Health Organization; that the economic elites control the world economy by inventing and operating the Federal Reserve Bank; that the Rockefellers are aiming to create a new world order; and that Charles Manson took part in a CIA experiment on mind control called MK-Ultra, which turned him into a serial killer.

When the Covid-19 pandemic first began, Richard believed what he heard on the news. Then he had an "awakening" and realized that he was trusting the wrong people (the "wrong people" being the CDC, the WHO, Bill Gates, Dr. Fauci, and so on). He began to "do his own research." He reads "all the scientific literature," he tells me, and when he doesn't understand something, he looks it up. Yet the result of all this is that he believes even more passionately in Covid-related conspiracy theories. How is he searching the vast universe of information out there and coming to his conclusions? It's a fascinating example of what social scientists call **biased search**.

Biased search, in the broadest sense, is any type of search that reveals only part of the universe of available information. Of course, it is hard to imagine a perfectly nonbiased search by any human being, so we should think about biased search as a continuum and question how biased the search is.

There are various influences that might bias our search. In some cases, it might just be that we look for information that is most readily available or easy to access—like the parable of the guy looking for his keys under a lamppost not because that was where he lost them but because that is where the light is. If we are searching online, the specific terms we enter, the algorithm the search engine uses, or the placement of sponsored results at the top biases the search. Perhaps we are using a search engine that unintentionally ignores certain data or language. Any search that is not fully representative is a form of biased search.

When it comes to misbelief, a particular type of biased search plays out, driven by what is known as **confirmation bias**. Confirmation bias is a biased search that starts with a hypothesis and looks only for information that supports that hypothesis, discounting or ignoring anything that might contradict it.

Richard exemplifies this. By the time I met him, he was convinced that Covid-19 was a tool used by governments to take away the basic liberties of their citizens. His extensive research was driven by the need to confirm that hypothesis.

In one of our online discussions, Richard sends me links to videos of people talking about the terrible effects the pandemic had on them, their relatives, the people they meet, and their kids. Some are clearly home recordings; others look more official. None of them strikes me as meeting any standard for evidence, and there is nothing much that I can say or learn from a video of someone describing how much pain they are having. I ask that he send me only hard data and academic papers. After all, he claims that he has been reading all the medical literature, so why not start there?

That night, Richard sends me numerous links accompanied by the statement "This information contains everything you need to know to understand what is going on." At first glance, many of them look like legitimate sources. The "experts" seem credentialed. The institutions seem real. The journals look like scientific medical journals.

I tend to be a very trusting person, but I know that in this situation I must go against my instincts and start every exploration by assuming that nothing is as it seems. Case in point: The first link I look at takes me to the Association of American Physicians and Surgeons (AAPS). It looks like any other medical institution's website with the standard logo of the rod of Asclepius (the snake around the rod), but a deeper dive reveals it to be a front for a right-wing political group that has little to do with medical science. Over the years AAPS has denied the existence of HIV/AIDS; promoted a link between abortions and breast cancer; asserted links between vaccinations and autism; claimed that homosexuality reduces life expectancy; and much more. To the untrained eye, the papers published on the site are formatted just like those in any other medical journal, but they are not based on any data or facts. They are opinion pieces and are full of speculation.

I reply to Richard, saying that this is not a scientific journal and I don't consider this the kind of evidence that I can trust. Richard doesn't even acknowledge my challenge; he simply moves to the next piece of information. This is typical of confirmation bias: he ignores any evidence that does not support his hypothesis. The next item he sends me is a tweet claiming that in Israel there was a 23 percent increase in mortality among men aged twenty-nine to thirty-nine during the months of March and April 2021. Because it coincided with the first wave of vaccinations for young people, he says, it is obvious that their deaths were caused by the vaccine. He also sends me a link to the Israeli Ministry of Health's online database. The data seem to be accurate: a 23 percent increase during these two months for this age group. However, a deeper look reveals the deaths in question to be a very small absolute number in the low double digits. When numbers are low, reporting increases as percentages can make them seem a lot more dramatic than they are. If two people die one week and then three people die the next, that's a 50 percent increase! And if, as in this case, the number of people dying increases from thirteen to sixteen, there is a 23 percent increase. By the way, this tactic of describing numbers in ways that are convenient to the person conveying the information is widely used, and not just by misbelievers. It is one of the main reasons that statistics sometimes gets a bad reputation.

Furthermore, the data were reporting aggregated deaths from a variety of causes. One bad car accident could have dramatically skewed the numbers. Next, I look more broadly at the mortality data, comparing this data point with a broader date range (not just April and March) and other age groups (not just twenty-nine to thirty-nine). It soon becomes clear that the numbers fluctuate over time and that the increase Richard is worried about is limited to this specific age group and those specific months. Take any other ranges of time and ages, and the increase is no longer visible. Next, I compare the increase statistic Richard sent me with the general fluctuations over time. I do this by looking at whether the increase (23 percent) is statistically indistinguishable from the distribution of general fluctuations of the number of people who die, and once again I find that although the 23 percent increase seems high, such fluctuations are expected and this specific increase is not statistically significant.

Hoping that this might be an opportunity to help Richard better understand science and statistics, I create a video of my computer screen as I am laying out my analysis and pointing to the different results. Again, true to form, Richard doesn't respond. He isn't interested in evidence that contradicts his beliefs; he only wants to find more evidence to back them up and further strengthen his conviction in what he already believes.

The truth is, we're all a bit like Richard. If you don't believe me, here's a simple test.

The Wason Cards

Imagine that you are presented with a set of four cards on a table that look something like this:

FIGURE 5

An illustration of the Wason selection task

You're told that each card has a letter on one side and a number on the other, and you're given the following hypothesis: *If a card shows an even number, then the opposite face shows a vowel.*

The question is: Which card or cards do you need to turn over to test the hypothesis? This is a test created in 1966 by Peter Wason, known as the Wason selection task.

Take a moment to think about it for yourself.

Most people will start with the card showing the number 4. If that card has a vowel on the other side, it will support the hypothesis. If it doesn't, it will disprove it. So that choice makes sense. Next, many people will pick the card showing the letter *E*. This seems like a very important way to test the hypothesis. But if you think about it, it is driven by a form of thinking that is looking to confirm the hypothesis (*If a card shows an even number, then the opposite face shows a vowel*). In reality, this is not a useful test. If the other side of the *E* is an even number, this will be consistent with the hypothesis, but if it is an odd number, it is not going to confirm or disconfirm the hypothesis because the hypothesis doesn't say anything about what should happen with cards that have an odd number on one side.

What very few people do in this test is to turn over the K card. But this card is actually the key, because if the *K* card shows an even number on the other side, it will disprove the hypothesis. The correct answer, therefore, is to turn over the 4 and the *K*. The other cards have no bearing on the hypothesis.

If you answered correctly, congratulations, you have an unusually logical mind, or else you have seen the Wason selection task before and are using this opportunity to confirm the hypothesis that you are much smarter than most people. If you answered incorrectly, you can be comforted by the fact that you are not alone. In Wason's original experiment, less than 10 percent of participants got the answer right. As you might expect, there is a large body of literature on the Wason selection task, with lots of nuances and conditions, and this literature by itself presents lots of opportunities for a biased search.

The Wason selection task demonstrates confirmation bias—the human tendency to look for evidence to support a hypothesis rather than to disprove it. Confirmation bias isn't limited to how we search for information; it also drives us to interpret information in a way that confirms or supports our preexisting beliefs or values and to avoid or ignore information that might challenge those beliefs and values. It enables us to quickly rationalize great inconsistencies in our conclusions about the world. The same person can insist that an election was stolen and the voting machines are unreliable if their preferred candidate loses but will praise the democratic process and have great confidence in the machines if their candidate wins. A person who believes in the weekly horoscope published in the newspaper can take every event that vaguely resembles the horoscope's predictions as proof that the stars have a plan for us based on when we were born, and that same person can also ignore or quickly explain away every event that doesn't fit the horoscope's predictions rather than question the horoscope's validity.

That's how Richard could read more into the meaning of the mortality statistics than was actually there and ignore all of my explanations. He discounted the high correlation between general vaccinations and the overall reduction of illness because it did not conform to his beliefs. At the same time, he overemphasized the weak relationship between vaccinations and the deaths of twenty-nine- to thirty-nine-year-olds during March and April 2021 because it provided a good fit with his existing mental model of what was happening in the world.

HOPEFULLY HELPFUL

Change the Way You Search

Search engines are a tool most of us use every day, but we often don't realize the degree to which they can support our confirmation bias. Have you noticed how, when you begin typing a query, the search engine offers ways to complete the query, affirming that lots of other people have that question as well? Have you also noticed a tendency in yourself, when searching for information related to a deeply and dearly held belief, to type in the thing you are trying to prove? If so, you're not the only one. Most of us do this. For example, if we believe that vaccines cause autism, we might search for the phrase "vaccines cause autism." This will lead us to content that confirms our belief. Too often, we think we are simply searching for a topic when we are really searching for our biased view of the topic. The flood of results we get further confirms our bias. I hope that one day, search engines will be designed to protect against confirmation bias. But until that day, we should do our best to counter it ourselves. Try typing in the opposite of what you believe. In the example above, that would mean searching for "vaccines do not cause autism." Doing both searches will lead to more nuanced and comprehensive search results and will guard, to some degree, against confirmation bias.

Confirmation bias gets all of us into trouble to some extent or another. Think about the last time you got into an argument with a friend and turned to Dr. Google for help. Did you look for all the information in a neutral way? Did you look up mostly (or only) supporting evidence for your point of view, or did you also seek to disprove it? Consider what happens when you hear a negative rumor about a person you don't like. Be honest: Do you approach the story with an open mind, checking the facts before drawing conclusions? Or do you assume that it *must* be true because it confirms what you already think about that person? You've no doubt encountered this in others as well. I do it, too (though not with horoscopes!). This tendency can be equally powerful when it comes to things we want to be true and things we are afraid may be true.

Searching for Cancer

I wake up and glance at my phone. It's two o'clock in the morning. As the fog of sleep recedes, I become aware of a strange sensation. The curve where my left shoulder meets my neck feels rather strange. When I touch that area, I realize that the sensation is muted and numb. I then draw a series of concentric circles, moving my fingers over a broader area each time, trying to understand the extent of the muted feeling and its boundaries. Eventually, I realize that there is a whole section of my body that feels oddly muffled. The feeling spans down my chest, up my shoulder, and down my back.

What could it be? My mind immediately says *cancer*. From there I spiral and think only about cancer-related questions. Where would a cancerous growth have to be located in my body to create this odd feeling? Would it be in my neck? My spinal cord? Could it be some kind of skin cancer? My thoughts take me to all kinds of bad places, so I decide to try to go back to sleep, hoping that the numbness will go away by the time I wake up. I turn over and eventually fall back asleep. When I wake up a few hours later, nothing has changed. The odd feeling remains, and so do the thoughts about cancer. I can't think about any other options to explain this feeling.

I become more certain that it's cancer and it is not going to end well. I spend the rest of the night in a fitful sleep full of nightmares. In the morning, the feeling is still the same. I know I should call a physician, but I don't even know what kind of cancer specialist to call. A skin cancer expert? An orthopedic surgeon? A spinal expert? Should I start with the emergency room or with my primary care doctor?

Then I touch my right shoulder, the non-cancerous one, and as one of my fingers touches the scars, it jogs my memory. Many years earlier, I had balloons implanted underneath the skin of both the right and the left shoulders as part of a surgical program to grow more skin and use the extra skin to replace some scar tissue (the balloons were originally designed for women's breast implants, but I suggested a new use for them, which my doctors first resisted but then embraced). Over the course of six months, the doctors slowly and methodically inflated each balloon. The one on my left shoulder was two liters in volume. The one on my right shoulder was a liter and a half. As they expanded, they stretched my healthy skin. The idea was that when the surgeons removed the balloons, they could stretch the extra healthy skin over my neck to replace some of the scars and give me a larger section of normal skin.

As I remember this, I touch the scar on my right shoulder and feel the same numbness there. At that moment, I remember that my shoulders have felt this weird way since that surgery, which was many years earlier. Mystery solved! This was an old problem that I simply forgot about. There is no cancer. The problem was that once I started down that bleak path, I obsessed about that possibility, and my mind could only produce more and more cancer-related thoughts. I discounted all other possibilities before I'd even considered them. I wish I would have stopped when I first felt this strange sensation and laid out *all* the options for things that could have muted the feeling in my shoulder. Instead I generated a single hypothesis, and all my thinking was constrained to focus on that one option.

The point of this story is that there are all kinds of ways to interpret things we experience, but once we set our mind in a certain direction (hypothesis), it is very hard to veer from that initial direction.

This same principle could explain some of what people claim are vaccine side effects. Imagine if the story I just told had occurred right after I'd received a vaccine. What would I have concluded? It's quite likely that I would have jumped to the conclusion that my numbness was a side effect of the vaccine—especially if it was in the same arm in which I'd gotten the shot. And if I was satisfied with that answer, maybe I would have not looked further. Indeed, when we expect side effects, we start looking for them. And sure enough, we are likely to find them. On any other day, the symptoms might not worry us; we'd remember what they came from, we would realize that they were random or stemmed from an altogether different cause. But when we expect side effects, we make them the focal point of our cognitive process. This can in turn enhance the magnitude of the symptoms we experience and create associations between those symptoms and whatever we expected to have caused them—in this case, the vaccine.

Connecting the Dots Backward

Speaking of creating associations, another common cognitive trick that reinforces misbelief is the tendency to find patterns of causation where none exists. We take two events that appear to be related but in fact are not, connect the dots backward, and then convince ourselves that A led to B. Let's take a look at one fascinating case of retroactively connecting the dots that many people engaged in during the Covid-19 era. It revolved around a 2019 pandemic scenario–planning exercise known as Event 201.

It couldn't be a coincidence, could it? At least, that's what Maya and many others would tell us. On October 18, 2019, the Bill & Melinda Gates Foundation and the World Economic Forum partnered with the Johns Hopkins Center for Health Security in New York to host a "tabletop exercise" modeling the possible responses to a pandemic. At this point in the book, just the list of the names of the players that were involved in this exercise is likely already raising your suspicions, maybe even raising the hair on your neck. You may be wondering if I attended, too (I did not), or if the Illuminati secretly sponsored the event (not to my knowledge). But this is not the end of the story. What kind of pandemic do you think they chose for their simulation? You guessed it: one caused by a coronavirus. It's hard to imagine how that could be a coincidence, right? What are the odds that they predicted Covid-19? Doesn't it make more sense that they already knew about it? Doesn't it make more sense to conclude that this "coincidence" proves that Bill Gates—seen by many as an architect of the pandemic—was already working on his evil plot and in cahoots with Johns Hopkins, another target of suspicion? The first cases of Covid-19 would not be publicly reported until December. Those seeking to confirm their hypothesis that Gates and his cronies planned it from the beginning have triumphantly seized on this event as proof. Let's admit it, the probability of having gamed out this scenario right before the outbreak of an actual pandemic seems statistically impossible if they didn't know what was coming. Right?

Wrong. This kind of thinking (confirmation bias) discounts the fact that such events were held regularly in the years prior to the Covid-19 outbreak. It discounts the rational explanations for the choice of a coronavirus as one of the more likely candidates for such an outbreak. It also discounts the fact that when searching among many events, it is easy to find the one event that in retrospect seems to fit a story we already believe. But what about the hundreds of other events that took place, many with the support of the Bill & Melinda Gates Foundation, that had nothing to do with the looming pandemic? Do we try to explain all of these? Of course not—no more than the horoscope devotee tries to explain all the events in a given week that were not predicted in last Saturday's column. Looking backward in time for chapters in a story we want to prove, it is easy to find some details that support our story. It's like looking first at the cards in the board game Clue and then guessing that it was Mr. Gates in the hospital with the needle!

Why Thinking of the Mind as a Computer
Gets Us into Trouble

Humans are drawn to think about everything in terms of metaphors; this is also true when it comes to how we think about our own minds. We look to the world around us for explanatory imagery, and since the world changes over time, so do what we see and the metaphors we use to reason about our own thinking. Plato famously compared the mind to an aviary; Vladimir Nabokov likened it to a filing cabinet; Sigmund Freud to an archaeological dig. The mind has often been described as a container of some sort, perhaps a leaky one. These days, however, one of the dominant metaphors is something that the ancient philosophers, poets, and psychologists could not even have conceptualized: the computer. With the proliferation of computers, we have started thinking about our minds as working like a computer. We use terms such as *bandwidth*, *processing power*, and so on when we talk about our own mental capacities. Though the computer is an interesting metaphor, like all metaphors it is also a limiting one. This limitation is especially clear when we think about the process of becoming a misbeliever.

The metaphor of the mind as a computer assumes that we are cognitive machines without any emotions and that our temporary emotional states don't change how our mind works. A computer obviously works as well when the stock market goes up as it does when the stock market goes down. A computer works just as well in the midst of a global pandemic as it does in times of extraordinary peace. A computer works just as well in the home of a family that is always in conflict as it does in the home of one that is in perfect harmony. A computer works just as well when the weather is terrible as it does when it is wonderful outside. This is not true of how people relate to information. The computer metaphor doesn't reflect the effect of stress, our motivations, or all the consequences of the fact that human beings are deeply social animals. Without taking these factors, among others, into consideration, the metaphor of the mind as a computer makes it harder—maybe even impossible—for us to understand and reason correctly about our own decisions and about the decisions of those around us. It simply leaves out all the truly amazing, quirky ways our minds work (and don't work). Which is why thinking about our minds in this way makes it hard for us to understand how someone could become a misbeliever. After all, it is very hard to imagine that a computer could become a misbeliever. So for the rest of our journey, let's try to not think about our minds and the minds around us as computers. Let's embrace the complexity and fallibility of the human mind and human existence.

Warning: Things Are About to Get Even Stranger . . .

None of the mental shortcuts I've described in this chapter, alone or in combination, guarantees that someone will become a misbeliever. But if we have stressful emotional conditions that drive us to seek relief in the form of answers, are driven by confirmation bias, and then seek and find misinformation that satisfies our need for a villain, it is likelier that we will form misbeliefs. The cognitive story is a complex one, and in the next chapter, we'll examine some even more convoluted ways in which we tell ourselves untrue stories and convince ourselves to believe them.

CHAPTER 6

Working Hard to Believe What We Already Believe

In the long run my observations have convinced me that some men, reasoning preposterously, first establish some conclusion in their minds which, either because of its being their own or because of their having received it from some person who has their entire confidence, impresses them so deeply that one finds it impossible ever to get it out of their heads. Such arguments in support of their fixed idea as they hit upon themselves or hear set forth by others, no matter how simple and stupid these may be, gain their instant acceptance and applause. On the other hand whatever is brought forward against it, however ingenious and conclusive, they receive with disdain or with hot rage—if indeed it does not make them ill. Beside themselves with passion, some of them would not be backward even about scheming to suppress and silence their adversaries.

—GALILEO GALILEI, *DIALOGUE CONCERNING THE TWO CHIEF WORLD SYSTEMS* (1632)

Have you ever tried to correct someone's misbelief by sending that person better and more reliable information? Maybe your aunt is convinced that AIDS was deliberately spread to wipe out the gay community, so you sent her some articles debunking that erroneous claim. Or some guy on your Facebook feed is declaring that the government possesses technology that can manipulate the weather, so you argue with him in the comments and share links to credible scientists explaining why it's not true. Or a colleague at work is terrified of 5G and has been warning you to stop using your cell phone before it's too late, so you print out some papers that should ease her fears and leave them on her desk. If you've tried anything like this, you'll know what I'm about to tell you: *It doesn't work.* Misinformation certainly feeds misbelief. But if that were the extent of the problem, we could easily correct it with good information. Unfortunately, this is not the case.

As we saw in the previous chapter, it is too easy to be drawn down a partial, biased, or simply false avenue of information as we search for what we already believe in. This is certainly an important factor in the creation of misbelievers. However, if the problem was limited to the way we search for information, the solution to misbelief would be as simple as improving our search patterns. We would simply need to expose our mind to a broader or different set of information and our subjective understanding of the world would quickly realign with the objective reality. Problem solved. But as you probably suspect, the human mind has additional forces that make it particularly difficult for us to be objective.

When someone wants to believe a particular narrative, it's extraordinary how the human capacity for reasoning can be harnessed in support of that story. Even when the story is blatantly false. In this chapter, we'll take a deeper dive into our cognitive structures, and examine the ways in which we more purposefully distort the evidence to reach the conclusions we want to reach. We'll also look at another, highly related key factor in the formation of misbelief, which is the gap between what we actually know and what we *think* we know: usually leading us to overconfidence.

But first, in case you thought I was talking about someone else when I referred to the purposeful distortion of evidence, let's take a look at an example you can probably relate to: sports.

Motivated Reasoning

Are you a sports fan? If so, feel free to substitute your favorite sport and team in the following example to make the emotional impact of this scenario more realistic. Let's take basketball. Say you're watching a high-stakes playoff game between your favorite NBA team and its most hated rival. Your team is leading by one point, with just seconds left on the clock. The archrival team has the ball and a chance to score the winning shot. Its best shooter steps back behind the three-point line, poised to attempt the shot. It's chaotic on the court as your team's defenders move in and players scuffle in the desperate final seconds. The shooter releases the ball, but it bounces off the rim of the basket and onto the court. A player from your team catches the ball. You're on your feet, about to celebrate, when a shrill whistle cuts through the roar of the crowd and the pounding of feet on hardwood. A foul has been called against your team. Your opponents will now shoot three free throws, and with next to no time left, they will most likely win the game.

From the stands, and from couches in countless living rooms across the country, a shout arises. Perhaps you're screaming yourself. "That wasn't a foul!" Fans of your team are yelling at their televisions, calling the referee all manner of names, accusing him of favoring the other team. The voices of logic from your side are brushing off the physical contact as "just part of the game." "Let 'em play!" fans are telling themselves and everyone else around them who is willing to listen. "Don't let the refs decide the game!" Even after watching the instant replays, again and again, they stick to their position, feeling robbed of an important victory by unfair officiating.

Now imagine that the scenario is reversed. The whistle blows, but the foul is called against the other team. Everything else is the same; the play in question is identical. How will the same fans react? How will you? Will you be suddenly sure it's a foul, able to explain exactly how the player breached the rules of the game? Will you feel confident that the referees are impartial and objective? Will you feel that they are fantastic at their job? Will you be happy that they stopped the play so close to the end of the game? Be honest. Most sports fans are guilty of this double standard: when a foul is called in their favor, it's all fair and clear, but when a foul is called against them, it's biased and poorly judged. And it's not just sports fans; it's human nature. We see things the way we want to see them, and we come up with very good arguments for the validity of our views. To put it in the terminology of cognitive science, we all practice **motivated reasoning**— the tendency to bend reality around us to fit the conclusion we want to reach.

We see motivated reasoning at play in many domains of life. Another very common one is politics. Back in 2016, when Donald Trump was up for election, it was amazing to see how many times he told blatant lies in a single day, and it was even more amazing to see how many Republicans didn't seem to care. Curious about that apparent infidelity to truth, I wondered if it was a defect of the Right or whether it was found on the Left as well. I decided to look into it. What I found was that the 2016 election was deeply ideological. People from both sides cared passionately about important topics such as healthcare, gun control, and abortion. It wasn't that the Republicans didn't care about honesty and dishonesty; it was that they cared much more about other issues. In fact, they cared so much about those other issues that they were willing to overlook a few lies if they served their higher aims. In addition, many Republicans actually looked at Trump's dishonesty as an indication of his commitment to their cause. Their reasoning went something like this: if he is willing to lie this freely now, it's likely that when he gets into office he will do whatever it takes to promote what is really important to our agenda.

Of course, when I told my friends on the Left what I'd found about why Republicans aren't more bothered by Trump's lies, they were appalled. But then I asked them the following question: Imagine there was a leader from the Democratic side who was so passionate about the issues of climate change, abortion rights, and gun control that they were willing to bend the truth a little bit in order to convince the people on the other side to pass laws that served their larger Democratic ideology. Would you be willing to support such a leader? Of course, that's a very uncomfortable thing to admit, but most of my liberal friends ended up admitting that because issues such as climate change, abortion rights, and gun control are so important, it would probably be okay to exaggerate and stretch the truth a bit—just for now, just until we pass the important legislation. And then we could return to the truth.

It doesn't matter if you identify with the Right or the Left; the point is that we all tend to see the world in the colors in which we want to see it. Our existing agendas shape our perception of how deserving our leaders are to be elected, their moral character, and the extent to which the end justifies the means.

When we think about honesty and dishonesty, there are not many people in the world who lie for the fun of it. But in many cases, there's a trade-off: Should we tell the truth, or should we pretend that we like a presentation (dress, meal, artwork, etc.)? Should we tell the truth, or should we promote our agenda? Should we tell the truth, or should we make a bit more money? Should we tell the truth, or should we keep ourselves in power? In all of these cases, it is not that people enjoy lying; it is that they try to convince themselves that something else is more important. In the hierarchy of needs, something else is more urgent right now, and when we lie it is because we have managed to convince ourselves that it's best to give up honesty in service of these more important objectives.

HOPEFULLY HELPFUL

Drop the Defensive Mindset and Be a Scout

In conversations with others, especially in difficult conversations, it's easy to default to a mindset of defending our territory. However, shifting out of that mindset can lead to a much more productive conversation. It is also useful to try to encourage the other person to make the same shift, but it's harder to change someone else's mindset than it is our own. Julia Galef refers to the posture of defending our territory as a "soldier mindset." According to Galef, adopting a soldier mindset in conversations entails viewing reasoning as a form of defensive combat, in which our values are at risk and we would do anything to defend them from opposing views. She suggests that we should adopt a "scout mindset" instead. Unlike the soldier, whose job it is to defend their territory, the scout's role is to explore and investigate. This requires having an open and inquisitive mind to properly map the territory. In other words, scouts are interested in what is true, what is out there in the world, whereas soldiers are hyperfocused on defeating threats. So always remember, if you want to avoid falling into misbelief, try to be a scout, not a soldier.

The Tale of an Unfortunate Doctor

In my personal interactions with misbelievers during Covid-19, there was one example of motivated reasoning that I found particularly fascinating. It involved the drug ivermectin and an unfortunate British doctor by the name of Andrew Hill who became the medical equivalent of the NBA referee in the scenario I described earlier in this chapter.

I first heard about this drug from Richard, who described ivermectin as a "Nobel Prize–winning drug" (which is true) and enthusiastically sent me a website link with lots of information about it. The link was connected to a larger website that collects studies on many "alternative" treatments for Covid-19, including aspirin, curcumin, diet, ivermectin, metformin, remdesivir, vitamins A, B, C, and D, zinc, and others. The information on the site declared that ivermectin was a cheap, safe, available, and highly effective treatment for Covid-19.

Looking at other online sources, I learned that ivermectin is a vital antiparasitic medicine used to treat mostly animals but also humans. In late 2020, as the pandemic continued to claim lives and vaccines had yet to be approved, many people hoped that this existing drug would be effective against Covid-19. Among them was Hill, a well-respected clinical researcher. That October, he was asked by the WHO to conduct a meta-analysis (a statistical analysis that combines the results of multiple scientific studies) of all the trials done so far on ivermectin and Covid-19. During that process, he tweeted several statements about the positive potential of ivermectin, earning himself the approval of the misbelievers, who by that point had seized on ivermectin as a potential alternative to the as-yet-unapproved vaccines. They saw Hill, with his medical credentials and his apparent enthusiasm for the drug, as the key to getting it approved as a Covid-19 treatment and thereby avoiding the need for a vaccine.

Hill and his scientific colleagues gathered the data from twenty-three trials of ivermectin from around the world and conducted a meta-analysis. The report was optimistic, noting that the results suggested that the drug could cut Covid-19 deaths by as much as 75 percent. But the researchers were also somewhat cautious and concluded that more data were needed in the form of larger randomized controlled trials before the drug could be considered for approval.

Those important points of caution were not reflected in Hill's brief Twitter statements as the research progressed, so when the full meta-analysis was made public on January 18, 2021, the ivermectin advocates were shocked and outraged. They accused him of sabotaging his own report with the caution of his final conclusion and betraying the cause to which they'd believed him to be loyal. It was as if the referee they had been convinced was in their corner suddenly reversed his call in the other team's favor. They needed a new story about

Dr. Hill, and they needed it fast. Whereas he'd previously been celebrated for his medical credentials and impeccable research, now he was denounced as a traitor and accused of putting his own selfish interests ahead of the welfare of humanity. His detractors became convinced that he was being paid off by Big Pharma, which wanted to discredit ivermectin in order to profit from vaccines and other novel drugs. They highlighted his links to the Gates Foundation and claimed that by delaying the approval of ivermectin, he was complicit in an "unconscionable" conspiracy that would lead to millions of unnecessary deaths.

In reality, Hill's meta-analysis was as supportive of ivermectin as any academic analysis could have been given the limited available data. Moreover, even with all the attacks, Hill remained positive about the outlook for the drug, calling the treatment potentially transformative. On the day after the report was published, he told the *Financial Times* that the conclusion of the meta-analysis should be to forewarn people: "Get prepared, get supplies, get ready to approve it." But for the ivermectin advocates, any delay amounted to outright betrayal.

Hill's popularity among misbelievers took a further hit in March 2021, when he posted a photo on Twitter showing him receiving his first Covid-19 vaccine shot. Horrified messages flooded in. Why would he do something like that? Why wouldn't he just use ivermectin? Was he on the Gates Foundation's payroll?

In the meantime, down the road at the University of London, a young medical student by the name of Jack Lawrence was assigned one of the key studies on ivermectin as part of his research for his master's thesis. That particular paper, by a professor in Egypt, was one of the most striking among the studies analyzed in Hill's report. It was the largest trial to date, with six hundred participants, and because of the large number of participants, it had a large impact on the meta-analysis. Stunningly, the results of the trial claimed that ivermectin led to a 90 percent mortality reduction. Lawrence wanted to get a closer look at the patient data, which were housed on a file-sharing site. He managed to hack the password and soon found trou bling issues with the data—such as subjects who had died before the study had begun. A couple of other "data sleuths" were also examining the data at the same time, and soon they found something even worse: lists of names that were repeated multiple times, as if someone had just cut and pasted the same patient data over and over again. It was clearly not data on which any conclusions should be based. The Egyptian study was not the only problematic study; closer investigation showed data problems in a few of the other ivermectin studies as well.

When that came to Hill's attention, he did what any scientist should do and revisited his original meta-analysis. He took out the studies he could no longer trust, including the Egyptian trial, and after removing those from the meta-analysis, he revised his conclusion: ivermectin provided no benefit.

His report of those findings in July 2021 triggered an even larger hate storm. Dr. Hill was bombarded with pictures of coffins, images of Nazi war criminals hanging from posts, and threats to himself and his family. The referee had not just called a foul on the wrong team, he had handed the game to the opposition. He must be not only corrupt but evil. Eventually, he had to shut down his Twitter account.

A few months after those events unfolded, larger randomized control trials confirmed the inefficacy of ivermectin for treating Covid-19. But none of that deterred those who were convinced of its lifesaving power. In fact, it all became part of the narrative: in the minds of the misbelievers, the pharma companies that stood to make billions of dollars from vaccines were purposely discrediting a cheap, safe alternative because it would cut into their profits.

In the misbelievers' online forums, they shared information on how to obtain the drug, and some of them even advocated using the versions intended for animals. They speculated that the world's elites (including the queen of England) must be secretly taking it when they became infected with Covid-19. When misbelievers became hospitalized with the virus themselves, they demanded the drug and became irate when it was refused. One man I met during my quest to understand misbelief even gifted me a full package of ivermec-

HOPEFULLY HELPFUL

Hesitate to Negate

When someone shares a piece of misinformation, it is very tempting to say "That thing you just said—*X*—is not true." This approach—repeating a statement and negating it—seems intuitively appealing, and in the short term it works. But the problem is that in the long term it can become less effective and sometimes can even backfire. Why? Because of two psychological quirks. The first is the **illusory truth effect**, wherein the more we encounter a piece of information (or misinformation), the more intensely the piece of information is coded in our brains as familiar and true, and the "stickier" it becomes. As you might have experienced, this particular quirk of human nature has not escaped the attention of advertisers and politicians. A strategy for taking advantage of this basic human quirk is attributed to Joseph Goebbels, who said: "Repeat a lie often enough and it becomes the truth." Or maybe Goebbels didn't actually say it but everyone thinks he did because of the illusory truth effect?

The second quirk is that because of the way our memory works, the two parts of the negating statement are stored separately in our minds. In the short term we code them as "*X*" and "*X* is not true," but over time, the link between them becomes weakened and this connection between the two statements loses its strength. As mentioned above, the illusory truth effect means that repetition is important, which is why the statement *X* ends up being judged as more true in our minds, while "*X* is not true" is largely discounted.

These observations give rise to two very clear recommendations. First, try not to let the illusory truth effect take hold from the start; this means that it is best to approach misbelievers with alternative and true information before they are exposed to a particular piece of misinformation, *X*, too many times. Second, we want to reduce our exposure to *X*. It is a much better approach to give an alternative story. Instead of saying "This thing you just said, *X*, is not true," say instead, "*Y* is true, and here is the evidence for it." Don't even mention *X*.

tin he'd purchased from an online supplier. Covid-19 was spreading quickly at the time, and he advised me to start taking the pills immediately in order to protect myself. I told him that I did not have much faith in that particular drug, but I could see that despite what I said, he believed that deep down I knew it was a wonderful medication and that with no one watching me, I would take it.

The story of ivermectin provides multiple examples of motivated reasoning. To start with, the proponents of ivermectin were hyperconcerned about possible flaws in the studies that showed ivermectin's ineffectiveness. They sounded very much like professional scientists as they pointed to a long list of potential methodological issues with each paper. They also questioned the ethics of the research team for every nonsupportive paper, bringing up all kinds of possible conflicts of interest. Yet they were unconcerned about other kinds of flaws, including plainly fraudulent data, in the papers that supported the use of ivermectin. They also had no problem completely revising their opinion about an individual such as Dr. Hill so long as it helped them avoid revising their opinion about ivermectin's effectiveness. They were masters at contorting the story by all means necessary in order to arrive at the conclusion they had in mind from the get-go. Their attitude was summed up succinctly by Dr. Tess Lawrie (not an infectious disease specialist but a specialist in pregnancy and childbirth), the founder of an ivermectin advocacy group. When asked what evidence might persuade her that the drug didn't work, she replied: "Ivermectin works. There's nothing that will persuade me."

Solution Aversion

When it comes to misbelief, there's a particular form of motivated reasoning that helps explain some of the more complex beliefs we commonly see in the world today. Here's a scenario that illustrates it. Imagine that there is a group of conservative Republicans, people whose political affiliations make them quite likely to believe that climate change is not a real issue. You give some of them an article to read that explains that solutions to climate change must include increased regulation and government intervention if we ever want to see a reduction in carbon dioxide emissions. Then you ask them "Is man-made climate change real?" and record their responses. You give others from the same group an article explaining that the solutions to climate change must include decreased regulation and government intervention and more free market–based initiatives if we ever want to see a reduction in carbon dioxide emissions. Then you ask them the same question, "Is man-made climate change real?," and record their responses. Do you think you'd see a difference in the way the two groups answered the question?

It turns out that the answer is yes. Troy Campbell and Aaron Kay, conducted an experiment along the lines of what I just described. They found that when asked how much they agreed with the scientific consensus that human-induced climate change would raise Earth's temperature by at least 3 degrees Fahrenheit during the twenty-first century, only a few of the self-identified Republicans who read the article focusing on regulatory solutions responded that they believed the statement was true (22 percent). Interestingly, the self-identified Republicans who read the article focusing on free market–based solutions were much more likely to say that they believed the statement was true (55 percent). What these results show is a phenomenon known as **solution aversion**. Put simply, it means that if we don't like a proposed solution to a problem, we use motivated reasoning to deny that the problem exists in the first place. Change the proposed solutions, as the researchers did in the climate change experiment, and suddenly we are more willing to admit that the problem is real. These results suggest that Republicans are not antiscience; it's that the standard solutions, which focus on regulatory controls and restrictions, conflict too strongly with their political values and ideology. Free market–based solutions create less conflict with their ideology and therefore trigger less denial.

In case you're thinking (or hoping, if you are a Democrat) that this is a Republican issue, the researchers performed a similar experiment with self-identified Democrats in order to make sure that it is clear to everyone that this human tendency goes beyond any particular political persuasion. This time, people were asked about the problem of violent crime, specifically "intruder violence" in the form of home invasions. The participants were given a series of questions about whether this was a significant issue. But before answering the questions, one group read an article focused on the idea that tighter gun control laws would prevent homeowners from being able to arm and defend themselves from intruders (an argument Democrats don't like). The other group read an article arguing that loose gun laws were making the problem worse, because intruders were more likely to be armed and cause more deaths (an argument Democrats like). After reading one of the articles, participants were asked to rate their agreement with a series of statements about the problem of intruder violence. As in the climate change experiment, the question was about the problem itself and not about the proposed solution. The results showed the same general pattern: Democrats (who generally support gun control) are more likely to deny or minimize the problem of intruder violence when presented with solutions favoring looser gun laws, which conflict with their political values and ideology. And they were more likely to rate the problem as severe when the proposed solution involved stricter gun control and therefore had a better fit with their ideology. Solution aversion, it seems, happens on both sides of the political divide.

Speaking of gun control, solution aversion related to this hot-button political issue is the key to understanding one of the more disturbing misbeliefs we've seen promoted in recent years. You've probably followed the case of the popular far-right radio host Alex Jones, who gained infamy by promoting and spreading a myriad of conspiracy theories, including the belief that school shootings such as the Sandy Hook Elementary School massacre were hoaxes staged by actors. That claim was so offensive to the grieving parents that they sued

Jones for defamation, and in 2022 were awarded more than a billion dollars. Many people watching the story unfold were dumbfounded. Why would Jones make such cruel and clearly ludicrous claims? And why would so many other people support and promote them, including people who have children of their own? Yet through the lens of solution aversion, it makes a twisted kind of sense. The number one solution that is proposed for the terrible problem of school shootings (at least by those on the political left) is stricter gun control laws. This solution is simply unacceptable to people (mostly on the political right) who believe that the right to bear arms is sacred and should never be taken away, no matter the cost. So because they don't like the solution, they deny the problem, even going so far as to call the tragic deaths of innocent children a hoax staged by gun control advocates.

Understanding solution aversion is critical if we are to make progress on some of the polarizing issues that continue to divide us. When liberals hear conservatives say that they don't believe in climate change, too often their instinct is just to push more information. "You didn't hear me. Here's another piece of evidence" is their basic response. And the liberals taking this informational approach are often deeply puzzled by the intractability of climate deniers' positions. What do these people have against data and science? How can they deny something that is a strongly established fact, at least in broad terms? How can they deny so much data? But what we find when we talk to climate change deniers is that they're not antiscience per se. What they don't like is the proposed solutions to climate change and the anticipated fallout of those solutions. They don't like the economic impacts of being asked to cut back on fossil fuels, for example, so they deny climate change instead. If we want to change people's opinions, we need to understand in more detail where their resistance is coming from. Often it is resistance to the solution, packaged in motivated reasoning, which means that until we come up with solutions that are more acceptable to both sides, one side will not give the information a chance. We often think that we should first agree on the facts and then move to figuring out the possible solutions, but when motivated reasoning is involved, we need to reverse this seemingly logical order and first deal with solution aversion.

Solution aversion can also show up in topics that are less controversial and more common in our daily lives. Let's say you are told by a doctor that you have a rare disease that is triggered by eating chocolate and therefore you must never eat chocolate again for the rest of your life in order to control the disease. How would you take this news? Would you believe the doctor? Or, faced with this undesirable "solution," would you be inclined to at least question, if not outright deny, the diagnosis? Be honest. You'd at least seek a second opinion, right? Fair enough. I probably would, too. But what would be driving us, in that moment, would primarily be our liking for chocolate and our dislike of the solution. If the doctor were to propose a different solution—a once-daily pill with no chocolate restrictions—we might be more inclined to accept his diagnosis.

Of course, we've seen this solution aversion dynamic play out on a large scale with Covid-19. On both sides of the political spectrum, we find people who are convinced that the virus itself is a hoax. People such

---

HOPEFULLY HELPFUL

Start with Better Solutions

Understanding how solution aversion works can be tremendously helpful in defusing conflicts and creating more productive dialogue. It reveals that our amazing ability to ignore or deny an obvious problem might really be resistance to the proposed, or even implied, solution. With this in mind, it is best to begin a dialogue by explicitly taking solutions off the table and trying to agree on the problem—for example, by highlighting and affirming that both sides are committed to the issue and care about mitigating it, even if they might go about it in different ways. An even better approach is to start the conversation by proposing solutions that will not be as objectionable to either side and, only when we have done away with fears of possible solutions, try to find common ground on the facts of the problem.

as those I've introduced in this book: Sara, Jenny, Brad, Richard, Maya, Eve, and many, many more like them. Are these people just science skeptics? Not necessarily. But the proposed solutions to the Covid-19 crisis were undesirable to them—both medically (in the form of vaccines) and socioeconomically (in the form of lockdowns and restriction of personal freedoms). So they took the next step in motivated reasoning and denied the problem. And no amount of evidence for the reality of Covid-19—whether in the form of scientific data, firsthand reports from people they know, or even getting the disease themselves—could convince them to change their view, because doing so would be tantamount to accepting a solution that was simply untenable to them.

How We Think and Misthink About Our Own Thinking

As we have seen in this chapter, the mind is a rather odd information-gathering and sense-making tool. One thing is for sure: it is not the objective tool we think it is. First, it does not measure everything in its path; it is biased and measures only some things and not others. Perhaps it is like a telescope that can observe stars very easily but not planets, presenting us with a biased view of the universe we are observing. Second, it is motivated to work hard to bend the world in order to fit our expectations. To continue with the telescope metaphor, imagine that its designers built a telescope that makes the observer feel as if it is scanning the skies smoothly but in fact it is designed to jump from one star to the next and avoid planets. This kind of design would be a very bad tool for gathering accurate information. However, the oddities of how the mind gathers information and makes sense of the world around us do not end there.

As previously discussed, a gap exists between the quality and objectivity of the mind and our perception of the quality and objectivity of the mind. It turns out that this gap is very important. Why? Because an imprecise tool is not a problem if we know that it's imprecise; we can take the imprecision into account. But if we have a bad tool or even a mediocre one, and we're unaware of its significant limitations, we'll keep using it as if it were a good tool and it may lead us astray.

For example, imagine a person who wants to measure the size of his living room but doesn't have a measuring tape. Luckily, he remembers that when he stretches out his hand, the distance from the tip of the pinky finger to the tip of the thumb is about nine inches. So he goes ahead and measures the dimensions of his living room. In the process, he becomes slightly distracted and is not sure if he counted to fourteen or fifteen, but he lets that go and continues from fifteen. When he has the final measurement of his living room, he knows very well that it is not an accurate measurement and consequently will not use it for any decisions that require precision. He'll use it only for rough estimates.

Unfortunately, when it comes to our minds, the story is very different. When we measure reality with our mind, we're almost always unaware of its imprecision. We don't understand how inaccurate our minds are, and because of that, we tend to trust them to a significant degree. The consequences can be as disastrous as a house that we designed and built using our hand as a measurement tool. Maybe the house would look interesting and creative, but it would certainly not be the house we wanted. And living in it would have serious consequences if there was a storm or the slightest earthquake. To use a more technical term, this is what **metacognition** is all about: the ways we think and misthink about our thinking and their consequences. Let's take a closer look at the relationship between our metacognition and misbelief.

The Dunning-Kruger Effect

Is there a gap between our actual knowledge and our confidence in our knowledge? It probably depends on the topic. For myself, if I take, for example, the field of physics, it is obvious to me that I know nothing about the topic. Therefore, my knowledge and my level of confidence in my knowledge are in general alignment. I read a little about physics from time to time, so maybe I think I know a bit more than I actually do. But in general, there's not a big gap between my objective knowledge and my assumptions about my knowledge. I will not go to a physics department and start arguing with a physics professor about the nuances of string theory. Now let's take an example from the other end of the spectrum: things I know a lot about. I happen to know a lot about the nature of dishonesty and the mechanisms by which people cheat a little bit and still feel comfortable with themselves. If I get into a discussion on this topic, I know that I know a lot. So here, too, there is no disconnect between my knowledge and my assumptions about my knowledge.

You may have noticed that in my examples, I skipped from a topic I know nothing about to a topic I know a lot about, without going into what's in the middle. What happens at this middle-knowledge level? It turns out that this is where trouble shows its complex face.

In social science, the trouble I'm referring to is called the **Dunning-Kruger effect**. It is based on the observation that our knowledge and our trust in our knowledge don't have to be the same. More specifically, this effect confirms what I've shared about myself: that when we don't know much about a particular topic, we often know that we don't know. And when we know a lot about a particular topic, we often know that we know a lot. But when we fall somewhere in the middle and know something about a topic but not that much, we often (mis)think that we know a lot more than we actually do. In these cases, there is a potentially dangerous gap between our actual knowledge and our confidence in our knowledge. These are the cases where we are ignorant but unaware of our own ignorance. Therefore, we can act with high confidence but are often wrong.

In my experience, this is what usually happens to college students in their first year of college. At the end of the first and second semesters, after they have taken the introductory courses in a certain subject, they think to themselves, "I have finished and mastered this course. After all, what else is there to know? I finished the whole introductory textbook, and I got an A on the exam. I feel I know this material from A to Z. I am the master of this material and maybe even a candidate to be the master of the universe."

I experienced that myself when I was an undergraduate and took a class on the physiology of the brain. When the semester ended, I'd finished the whole book, I'd gotten good grades, and I felt that I had a deep understanding of brain physiology. I was under the impression that I'd mastered the material. If I'd been asked to give myself a score at that point on how much I knew, I would have given myself a 90 on a 100-point scale. Since then, I have not chosen to specialize in the physiology of the brain, but from time to time I read articles about the topic or get involved in a research project related to it. Every paper I read and every research project I get involved with has increased my objective knowledge of brain physiology, but with every passing year since I was an undergrad, it has become clearer to me how little I know about the topic. In fact, my confidence in my knowledge of brain physiology was at its peak when I'd finished that first-year undergraduate course. Since then, my confidence in my knowledge has decreased systematically, year after year. My overconfidence when I had just completed the introductory course is the basic mismatch that creates the Dunning-Kruger effect.

Now, if the Dunning-Kruger effect were limited to our own self-knowledge, it wouldn't be so bad. But it doesn't end there. Why? Because the gap between objective knowledge and confidence opens the door to very undesirable consequences. It basically creates overconfidence. Somebody who thinks they understand how motivation works more than they actually do can start a company and design an incentive plan to increase employees' motivation but end up decreasing their motivation instead. Somebody who thinks they understand behavioral change when in fact they know very little about it can create an app designed to help people lose weight but in fact have very little impact. And somebody who believes that they understand how viruses and the immune system work, when in fact they don't, might make all kinds of decisions about vaccinations, medications, and treatments that are not supported by the evidence.

The Dunning-Kruger Effect confirms what the poet Alexander Pope once wrote: "A little learning is a dangerous thing." We might not think that applies to us because we know more than a little, right? Or do we? Do we really know as much as we think we know?

The Illusion of Explanatory Depth

Another psychological quirk related to the Dunning-Kruger effect is called the **illusion of explanatory depth**. Before we get into more detail, let's start with an exercise, based on the work of Rebecca Lawson.

Bicycle Knowledge

Please mark your answers to each of the questions below. And again, please do not just answer them in your mind.

1. Have you ever seen a bicycle before? Yes/No

2. Do you know how to use a bicycle? Yes/No

3. Do you understand how a bicycle works? Yes/No

4. How well do you understand how a bicycle works?
   (using a scale from 0 = not sure at all, to 100 =
   understand it perfectly well): _____

5. Look at the bicycle below and think about what a
   real bicycle should look like. Then draw the frame,
   pedals, and chain in their correct positions.

6. Now let's look at the same question but focus on one part at
   a time: Looking at the four images below, which bicycle shows
   the usual position of the frame? (please circle your answer)

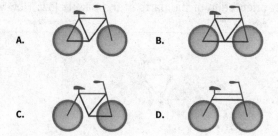

7. Looking at the four images below, which bicycle best shows the usual position of the pedals? (please circle your answer)

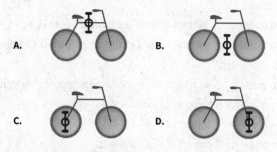

8. Looking at the four images below, which bicycle best shows the usual position of the chain? (please circle your answer)

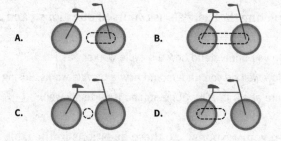

See the correct orientation of the parts of the bike below. How well did you do? _____

Now that you have an example of how a bicycle really looks, add up your score:

In your answer to question 5, is the frame in the correct spot? (1 point if it is)
In your answer to question 5, are the pedals in the correct spot? (1 point if they are)
In your answer to question 5, is the chain in the correct spot? (1 point if it is)
Is your answer to question 6 correct? (1 point if it is)
Is your answer to question 7 correct? (1 point if it is)
Is your answer to question 8 correct? (1 point if it is)

Total Score: _____/6 points

After completing the exercise, ask yourself questions 3 and 4 again:

3 (Updated). Do you understand how a bicycle works? Yes/No
4 (Updated). How well do you understand how a bicycle works? (using a scale from 0 = not sure at all, to 100 = understand it perfectly well): _____

Write down your responses to these questions in the table below:

	Response Before Exercise	Response After Exercise
3. Understanding (Yes/No)		
4. Understanding (0–100)		

To what extent did you change your mind? (using a scale from 0 = not at all, to 100 = very much) _____

To what extent do you think there are other things in your daily environment that you thought you understood but now you are not so sure? (using a scale from 0 = there must be many other things that I overestimate how much I understand, to 100 = I don't think there is anything else that I overestimate how much I understand) _____

You may have found the test above revealing. Or maybe not. Perhaps you are an expert bicycle mechanic. Perhaps you enlisted the help of the internet. Perhaps you made a detour to your garage to remind yourself what bicycles look like. So let's take a look at the results of a real experiment that was carried out under controlled conditions, using the same basic approach.

Leonid Rozenblit and Frank Keil set out to explore the illusion of explanatory depth: the human tendency to intuitively feel that we understand complex phenomena much more deeply and precisely than we actually do. To begin, the researchers gave participants a list of forty-eight common objects, processes, and natural phenomena ranging from can openers to livers to photocopiers to presidential elections. The participants were asked to go through the list and rate their level of understanding of each on a scale of 1 to 7, 1 being the lowest level of understanding and 7 being the highest. Then they were asked to take four of those items (selected by the researchers) and write a detailed, step-by-step causal explanation of how they worked. For example, a participant might be asked to give a detailed description of how a flush toilet, a helicopter, or a sewing machine works.

After writing down their descriptions, participants went back and revised their rating of how well they understood the particular items they had just tried to explain in detail (much as you were asked to revisit your understanding of bicycles). Then the researchers took it a step further: they asked the participants a "diagnostic" question related to each of the same four items—the kind of question that they

---

HOPEFULLY HELPFUL

Paradoxical Persuasion

Given the amazing flexibility of the cognitive system and our ability to quickly defend ourselves against incoming information and disarm it, what type of persuasion might have an impact? One effective approach is that of overly agreeing with the person you want to persuade—not just agreeing but taking a more extreme perspective than they do. Often, as we've seen, people confidently adopt a position and assume that they understand its implications without really thinking it through. When you take their suggestion even more seriously than they take it themselves, you expose the shallowness of their position. Someone tells you that all pharma companies are evil? Agree with them. Then suggest that they should stop taking all medications and cancel their medical insurance immediately. Someone tells you that 5G is dangerous? Agree with them. Suggest, in fact, that they might get rid of their cell phone altogether and rely on landlines. Or maybe that's not enough. Perhaps they should consider moving to Green Bank, Virginia, a town where there is no Wi-Fi and cell phones are banned in order to protect the city's resident giant telescope. Pull up some real estate listings for them. Sure, there would be downsides to living in a tech-free zone, but it would be worth it, right, if 5G is going to kill us all?

This approach turns out to be rather effective at making people reconsider their extreme positions.

---

could answer correctly only if they really understood the mechanism. If the item was a helicopter, for example, they asked the participants to explain how it shifts from a hover to a forward flight. After answering this additional question, the participants were asked again to revise their rating of their understanding. Next the participants were given descriptions of the items written by experts, and after reading those reports they were asked to revisit their rating one last time.

The results show that people start out feeling quite confident in their knowledge but their confidence drops after they are asked to give a detailed explanation and it drops even further after they answer the diagnostic question. However, their confidence stays the same (low but the same) after reading expert descriptions.

The first part of these results shows that we generally think that we know more than we really do, to such an extent that even just thinking more concretely prompts us to recognize how little we know. It would be one thing if, by learning new things, we figured out that we don't know as much as we thought we did, but the first part of the experiment on the illusion of explanatory depth purposefully did not provide the participants with any new information. They were simply asked to reflect more explicitly and concretely on their knowledge and understanding; that alone was sufficient to create a better match between what they really knew and what they thought they knew.

The second part of the results, the finding that there is no change after being exposed to the expert descriptions, shows that by the time the participants got to that stage, their low perception of their own knowledge was basically accurate and could not go any lower, even when they were confronted with the knowledge of an expert.

Another interesting finding from that study was that the degree of overconfidence is not the same for all items and that it is particularly high for items that have a higher ratio of hidden to visible parts (such as a computer). For these types of items, we can more easily hold the belief that we understand the object and therefore are more likely to be overconfident. Now think about what this means for devices, items, phenomena, or processes that are invisible, such as viruses, the immune system, vaccinations, global warming, 5G chips, and so on. When it came to Covid-19, almost everything was invisible, so people's overconfidence levels, their illusion of explanatory depth, and their ability to maintain their confidence was likely very high.

Inspired by that experiment, I decided to conduct my own study, involving a flush toilet. Imagine that you are one of the participants. You show up at the lab, and a friendly person in a white coat asks you if you know how a toilet works. "Sure," you reply, and go on to rate

HOPEFULLY HELPFUL

Challenge the Illusion of Explanatory Depth at Home

Sit down with Uncle Joe and ask him to explain in depth how the mechanism of his pet misbelief works. How exactly is 5G changing people's cells, and how is it different from a microwave in the kitchen? How is the United Nations, an organization that could not stop the war in *X* (put your favorite sample here), able to control the world? How exactly is the government controlling the weather? The point is to try to follow the recipe for the illusion of explanatory depth test and get Uncle Joe to realize that he does not know as much as he thought he did about what he believed before you started talking.

What if Uncle Joe is unwilling to discuss his pet misbelief? Or what if his misbelief is such a hot-button issue that you don't feel comfortable bringing it up in the short time you have together? Or what if the last time that you brought up the topic the argument was so painful that you just don't have the energy to go down that path again? Interestingly, it seems that there is a way to manage such sensitivities and diffuse the illusion of explanatory depth. As Ethan Meyers and his colleagues demonstrated, reducing the illusion of explanatory depth in one domain can help reduce it in other domains as well. For example, they showed that after proving to people that they didn't really know how a zipper worked, those same people also started doubting their own knowledge of how snow is formed. And although Meyers and his colleagues did not test the extent to which such transference of healthy self-doubt can occur for very strongly and centrally held beliefs, as is often the case with misbelief, their results suggest that this is possible. So if the discussion with Uncle Joe about his pet misbelief is not very promising, it is okay to start with his understanding of more mundane and less controversial topics—such as how his hoodie stays zipped.

your understanding on a scale of 1 to 7. (The participants, on average, rated themselves at 5.1.) Next you're given a paper and pencil and asked to draw a diagram showing how a toilet works. Once you're done, you are given the option to revise your rating. Our participants generally felt that they had been slightly overconfident, and after struggling to draw the toilet, they dropped their rating, to an average of 4.2. Next you are led to a table on which all the various parts of a toilet are laid out. We assure you that they are new and clean and ask you to reassemble the toilet. That was an even greater challenge for our participants than the drawing was. After playing with the parts for a while and trying to assemble the toilet (no one was able to do it successfully), their average rating of their knowledge dropped to 2.7.

You may be wondering if it really matters that you don't know how a toilet functions. Or a zipper. Or a sewing machine. Of course, your life is unlikely to be negatively affected if you don't fully understand how many of these items work, but the point is to shine a light on the more general problem of overconfidence. The gap between what we know and what we think we know can be dangerous, even life threatening. One such example from my own life involves the ability to drive. At a certain point, I took a two-day driving course, which left me feeling extremely confident in my driving ability. Within the following two weeks, I had a car accident, which I'm fairly sure was a result of my overconfidence, courtesy of the course I'd taken. When we make decisions that are based not on our real knowledge or skill level but on our perceived knowledge or skill level, a dangerous gap is created—a gap into which we can fall if we're not careful.

Persuasion Doesn't Always Work

The insights we get from the research on the illusion of explanatory depth makes me think about Maya, whom we met in the previous chapter, and the way she shared the magnet conspiracy theory with me. I'm pretty sure that if you asked Maya if she understands how a magnet works, she would give herself a very high score. She might even claim to understand the mysterious-sounding magneto-fection. How can I be so sure? Especially in a chapter that is focused on people being too confident about their own knowledge? Well, in this case it is because I actually talked with Maya about it and tried to challenge her description using basic physics. As I suspected, she revealed herself to be much more confident than her actual level of knowledge would justify. Unfortunately, though reasonable people tend to revise their confidence levels in experiments like the ones described above, misbelievers such as Maya usually don't. Their reasoning, as we have discussed, is motivated by their beliefs, and this makes their overconfidence much more intractable.

I began by asking Maya if she knew how a magnet works. Her response showed modest confidence in her own knowledge. Then I challenged her with a few facts, including that it would take a very large magnet to hold an iPhone or even a coin, much larger than could fit inside a vaccination needle. I also informed her that the US coins shown in some of the videos are made of a nonmagnetic metal. Plus, I pointed out, "Some of the videos are inconsistent with other videos. The magnetic force can be either on the inside of the body or the outside, but it can't be in both places." If Maya had been a participant in an experiment testing the illusion of explanatory depth, at that point in the process I would have expected that being reminded about how magnets work might dramatically decrease how she rated her knowledge. But she was not interested in facts, so instead, she simply pivoted to social proof again. "I've seen it myself," she declared. "My friend shared pictures of utensils magnetized to her body."

Maya's ability to hold on to her beliefs despite the very clear evidence against this particular theory (and it is rather rare to encounter a conspiracy theory that is so easy to refute) suggests that if there were a boxing match between two legendary fighters—in one corner the ability to rethink and combat the illusion of explanatory depth and in the other corner motivated reasoning—motivated reasoning would win hands down.

Realizing that, I changed direction and asked Maya if she could connect me to her friend so that I could see the "evidence" in person.

She promised to do so but never did. Maybe she realized that I was so stubborn in sticking to my misconceived ideas—despite clear video evidence and a report from Stew Peters and his impeccably credentialed doctor guest—that there was no reason to waste more time trying to convince me.

Overconfidence and Fake News

Earlier in the book when I described *The Stew Peters Show* or the Association of American Physicians and Surgeons' website, you likely thought that you would easily be able to tell that they were untrustworthy sources. And you might be right. Or you might be exhibiting overconfidence. It's hard to say in your particular case, but it turns out that overconfidence is a big problem when it comes to fake news. A group of researchers set out to understand the relationship between overconfidence in news judgment and susceptibility to fake news. Benjamin Lyons and his colleagues conducted large-scale surveys that revealed disturbing results. First, they found that 75 percent of Americans overestimate their ability to detect fake news (on average by 22 percent higher than warranted). Moreover, they found that those overconfident individuals were more likely to fail when asked to distinguish between true and false claims about current events, providing further evidence of the relationship between lower knowledge and higher overconfidence. Even worse, they found that those with lower factual knowledge were more willing to share false content on social media. In other words, those who are unknowledgeable are the most overconfident in their judgment of the news, are more susceptible to fake news, and are more likely to spread it further.

In Summary . . .

The interconnected story of the human mind and the information it seeks and processes is anything but simple. We tend to think that we are information machines that objectively take in information, re-cord it, analyze it, and come to logical conclusions. But nothing could be farther from the truth. We have evolved a complex, sometimes sophisticated, sometimes flawed, and sometimes downright faulty set of shortcuts for processing the overwhelming world we live in. These sense-making mechanisms serve us well—until they don't. In today's world, beset by all the stresses we've already examined, we find ourselves confronting a vast sea of information and misinfor-mation, delivered to us through channels that sometimes try to take advantage of our biases without our knowledge. And that's not all. The human mind has an extraordinary capacity to create a narrative that fits the conclusion we want to reach. If we don't like a conclusion, we rewrite the narrative.

Furthermore, the world is complex, yet as human beings we are of-ten satisfied with overly simple explanations. We don't process infor-mation very deeply, and for most of us, the *feeling* that we understand something is sufficient. All of these factors contribute to the creation of misbelief. When compounded by emotional challenges such as unpredictable stress and the feeling that we don't have control, they can create powerful convictions that are hard to dislodge. Does that mean anyone who thinks in these ways becomes a misbeliever? Not necessarily. As we will soon see, personality plays an important role. Some personality traits make people more susceptible to the forces of misbelief. We will take a closer look at these traits in the following two chapters. Then we'll look at how misbelief is deepened and rein-forced by misbelievers' social circles and the social forces that keep people loyal to conspiracy theories and the communities that spread them.

FIGURE 6

The cognitive elements of the funnel of misbelief

··· When stress pushes us down the funnel and we begin to search for answers and look for a villain to blame, cognitive elements lead us deeper into misbelief.

··· Faulty human cognitive structures make us susceptible to misinformation, just as our evolutionary predilections make us susceptible to fast food.

··· When we search for information, confirmation bias leads us to look for things that confirm our suspicions rather than disprove them.

··· Once we believe something, we work hard to convince ourselves that it's true—a process known as motivated reasoning. Sometimes, our fear of implied solutions to problems leads us to deny the problems themselves—a bias known as solution aversion.

··· Conspiracy theories are designed to take advantage of our cognitive biases.

··· All of this is exacerbated by the ways in which we misthink about our own thinking and fall prey to overconfidence in our understanding of how things work.

PART IV

THE PERSONALITY ELEMENTS AND THE STORY OF OUR INDIVIDUAL DIFFERENCES

CHAPTER 7

Lessons on Personality from Alien Abductees

We meet aliens every day who have something to give us.
They come in the form of people with different opinions.

—WILLIAM SHATNER

A Lost Saul

I never met Saul, and unfortunately I never will. But at some point in the last few years, he had appeared so often on my Facebook feed that I started following him. Picture a stocky man in his sixties, balding on top, recording himself in selfie mode while walking outside his home, sharing with his real and imagined audience his views and reflections about what was *really* happening in the world. Many mornings, I tuned in, and I grew to feel a certain affection for Saul—the kind of affection we might have for a retired neighbor who strolls past our house every day, well meaning and kind but with a little too much free time to lean over the fence and share his philosophy of life.

Among my personal pantheon of misbelievers, Saul was one of the more likable characters, clearly motivated by compassion and a desire to help people, however misguided. He styled himself as a kind of wise tribal elder, adopting an air of self-importance as if he was trying hard to convey to us, his followers, that he was someone to whom we should pay close attention. Judging by the number of views of his videos, a lot of people did. And I did, too—though not for the reasons he would have wanted.

One particular morning, Saul's voice took on an extra tone of seriousness as he delivered his daily speech. His topic was Event 201, which we discussed in chapter 5. You may remember that in November 2019, Johns Hopkins University, the World Economic Forum, and the Gates Foundation were all involved in a pandemic simulation exercise featuring a coronavirus. What more proof could you need, Saul asked, that Covid-19 was a plandemic (planned pandemic)? Could it be a coincidence? Obviously not. For Saul, this was unassailable proof that the plandemic had been set into motion by those evil actors.

Over time, Saul began recording his selfie videos at demonstrations. He'd go on Facebook Live while protesting Covid-related restrictions or vaccines. From demonstration to demonstration, he became more belligerent—first verbally abusive and then physically violent toward the police. He began to seem less like a harmless rambling neighbor and more like the ranting guy you cross the street to avoid. Watching his progression over those months, I often wondered: What happened to Saul? How did he get here? It was clear to me that he must have lost something very significant in the early months of the Covid-19 pandemic and now he was looking for attention and social connection. From his recorded morning talks, his live reporting during the demonstrations, and the increasing intensity with which he propelled himself into a leadership role among the misbelievers, his need for attention was palpable.

At some point, Saul got a stamp of approval from an unexpected source: he was arrested for his violent behavior at a demonstration. After he spent the night in jail, it seemed as if he had hit the jackpot. He was now a bona fide outlaw with an official record. After that, it looked as if he was trying to get arrested at every demonstration, and always on camera. Each time he was released after a night in jail, he reported on how the police had mistreated him. He seemed to view himself as a kind of Nelson Mandela figure, fighting for freedom and human rights against the globalist cabal.

Saul kept reporting, and I kept watching. In yet another one of his morning-walk recordings, he commented on recent reports of an increase in hospitalization rates among older individuals. That was not fake news, he declared. It was real. But, he explained, the reports cited a misleading reason for the increased hospitalizations. And, he added, the media knew it. So what was going on? Saul quoted a report showing that since the beginning of Covid-19 the use of technology by older individuals had been on the rise. This increased use of technology, he claimed, was the actual cause of the rise in hospitalizations. His point was that due to their weaker immune systems, the older population was more susceptible to the harmful radiation from technological gadgets. In other words, the increased hospitalization among elders was the fault of the government restrictions and greater use of technology and had nothing to do with the plandemic.

One day, I clicked on Saul's video and was surprised to see that he was not on one of his usual walks or attending a demonstration. Instead, he was sitting in a hospital bed, connected to an oxygen line, his breathing clearly labored. Still, he was reporting live to his devoted followers. He told us that he had tested positive for Covid-19 and was now hospitalized in a Covid ward. But, he insisted, he didn't have Covid. He'd been arrested at a demonstration a few days earlier and was certain that the police had poisoned him while he was in jail. He described the effects of the poison, including great difficulty breathing. His poisoning, he concluded, must mean that the police were getting desperate. In closing, he begged his followers to keep on fighting and not stop their important work.

Saul died a few days later in that same Covid ward. I am not sure how many of his followers went to his funeral or how welcomed they were by his grieving family. But his Facebook page was filled with sentiments of love and admiration and promises to demand an investigation into his poisoning.

How could Saul have maintained his beliefs in the face of so much counterevidence? Even when he tested positive for Covid-19 and could barely breathe? How could someone who had appeared to be such a gentle soul have become so verbally abusive and later physically aggressive to members of the police force? What were the forces that drove him to connect the dots in such a creative yet misguided way? What caused him to become a misbeliever in the first place and to maintain his conviction with such strength?

Of course, it is hard to know what was really going on in Saul's mind, and due to his untimely passing, we will never be sure. But observing him for so many hours suggested to me that his personality played an important role in the process. For one, there was his constant need for recognition and approval from others, made evident by his endless Facebook Lives, by his escalating violence over time, and by the jailed martyr role he adopted to gain influence among misbelievers. There were also his ongoing monologues that displayed a propensity to connect unrelated and nonexistent dots. And of course, there was his lack of critical thinking that was most obvious when he was diagnosed with Covid-19 but refused to see it. Could those traits account for his descent into misbelief? Not by themselves. But they provide another important perspective from which to understand what makes some people more susceptible to the funnel of misbelief.

I would guess that Saul—like Sara, Jenny, Brad, Richard, Eve, Maya, and the rest of us, to some degree—experienced unusual and unpredictable stress in the early 2020s. Saul then began searching for answers and for someone to blame. With the aid of his cognitive biases and his particular combination of personality traits, his path down the funnel of misbelief was well lubricated. In this part of the book, we'll take a deeper look at the role personality plays and the ways it primes us for misbelief. Specifically, in this chapter we will try to understand the delicate interaction between personality structure and susceptibility to misbelief.

Some General Words About Personality

Before we dive into the types of personalities that are more likely to usher someone down the funnel of misbelief, a few words in general about personality. Put most simply, personality refers to the particular combination of traits that makes up a person's distinctive character and the essential quality that makes each of us who we are. It's like adding splashes of color to a black-and-white photo. So far in our journey, we have focused mostly on the ways in which we are all similar, and there are indeed many similar aspects of our humanity, from our emotional responses to stress to our cognitive biases. However, it is important to acknowledge that we are not exactly alike and there are differences in how different people react to the same situation and the same information.

For example, in our discussion of stress, it was clear that personality structure can have an impact on our experience of stress and hence on our susceptibility to stress. Two people can be exposed to the same stressors and react differently. Just try to put yourself into my shoes and imagine how you would react if you were to find yourself the target of online hate and conspiracy theories. Would you feel it more intensely than I did, or would you be more lighthearted about it? Let's hope that you never find out. But try to imagine it, and then compare how you think you would react to the way your significant other or one of your good friends would react. Most likely you will get the sense that although stress is bad for all of us, the extent of its impact varies by individual. Hence, we can think of stress as a significant external force. We can also think of susceptibility to stress as an important individual difference. The same is true of the need to make sense of the world. This sense-making motivation is different in its intensity for different individuals, and since it is an important element in the funnel of misbelief, it may partially account for why some people are more susceptible than others.

There are times in our personal and general history (loss of a job, illness in the family, Covid-19) when any of us would become more susceptible to the harmful effects of stress. Jenny, for example, was experiencing both the general stress of her environment and the specific stress of her personal situation (both work and family). This stress was exacerbated by her particular personality, leading her to start her journey down the funnel of misbelief.

Obviously, personality is far more complex and far more interesting than just susceptibility to stress, so in this chapter and the next, we will try to uncover some of the complexities that come with personality.

Decoding Personality: Traits and States

In the examples cited above, there are two types of personality constructs: personality traits and personality states. **Personality traits** are what we usually refer to when we describe someone's personality: Sally has a higher level of empathy; Bill is more of a narcissist; Julio is stingy. My rambling Facebook reporter Saul's particular mixture of personality traits included some degree of narcissism, a tendency to see patterns where none exist, and an ability to connect dots creatively. Such personality traits are individual differences that people such as Saul carry with them wherever they go. Traits are relatively stable and consistent over time, and therefore they become a lens through which people experience everything in their life.

However, you don't need a degree in psychology to know that people aren't always consistent. Sometimes we act in ways that seem "out of character," at odds with the traits we display most of the time. This is where the concept of **personality states** is a better way to describe people. States are temporary changes in personality that occur in particular circumstances and change us, for a short time, in substantial ways. Think about road rage. Perhaps you have experienced this at some point. You were calmly driving down the street with your loved ones when all of a sudden someone blazed past you at a dangerous speed, cutting you off and putting you and your passengers at risk. I am willing to bet that at that moment you became a slightly (or maybe not so slightly) different person. If the reckless car had stopped next to you at the next traffic light, you would have probably shared with the driver your opinions about them and their ancestors in not-so-flattering language. Those of us who have experienced road rage are almost always amazed by who we became in those moments. And our family members are even more surprised. Don't believe me?

Go and ask one of your family members or close friends about an episode in which you were enraged in this way.

This type of short-term change is an example of a personality state—a situation in which we temporarily become a different person. If you really can't relate to the road rage example because you always maintain perfect equanimity behind the wheel of your car, think about a time when you were deeply in love. No doubt, in the midst of that intoxication, you did things that seemed quite different from your everyday personality. Perhaps you were suddenly willing to engage in public displays of affection when you are generally more reserved. Or maybe you became extravagantly generous, buying expensive gifts for your beloved, when your normal personality spends more conservatively. Similar shifts can occur during a sudden emergency or natural disaster. Normally timid people may exhibit unusual courage, and people who tend to be passive or inclined to follow can suddenly step up and assert themselves as leaders. In all of these Dr. Jekyll and Mr. Hyde examples, we see that changes in the environment can temporarily change us in ways that are powerful enough to seem "out of character."

The personality-related topics touched upon in earlier chapters were mostly from the state category. This chapter will deal mostly, but not exclusively, with personality traits. To be clear, these two personality constructs are intimately linked. For example, if we find that there are people who are more sensitive to stress (as a trait) and because of that are more likely to go down the funnel of misbelief, this would also suggest that temporary stress (the kind that can create a personality state change) is likely to play a role in their descent. Conversely, if we find that temporary stress changes the way people go down the funnel of misbelief, it is also likely that people who are sensitive to stress as a trait would be influenced to a greater degree.

Another point worth mentioning concerns the way we interpret the notion of personality. When we hear that an individual has personality X or Y, it is tempting to interpret that as meaning that this personality creates the person's general behavior in many different circumstances. Someone who is generous will be generous anywhere they have the chance. Someone who is creative will express creativity in every activity. Someone who is a narcissist will exhibit self-importance everywhere they go. Not so! Personality is like a small side current in a flowing river. Yes, it can help push objects in a certain direction, but it is only a part of the whole picture. In fact, it is small in magnitude compared with the overall current of the river, and it will act in different ways on different objects, depending on their weight, buoyancy, shape, and so on. For example, I once tested extreme-sport athletes—the kind of people who jump off mountains on bikes, climb precarious rock faces, or fly in those body suits that make them look like flying squirrels. In short, real daredevils. It was obvious that those people took a lot of risks in their athletic pursuits. You might conclude that their personality traits included high risk taking. But were they risk takers in general? Did their appetite for risk extend to other types of risk? In the stock market? In their work life? In their personal life? The answer, I found, was no. In those other areas they were plain vanilla risk takers. The point is that although they were certainly risk takers, they were not risk takers in every arena.

What does this mean when it comes to the role of personality in misbelief? Well, as with each of the factors we've considered so far, it's complicated. Just because someone has a high degree of narcissism (I mention narcissism a lot because, as you will soon see, it is strongly connected with misbelief), that doesn't mean that they will necessarily tumble down the funnel of misbelief. Nor is someone who is low on the narcissism scale going to be fully protected from this path. At the end of the day, no single personality trait or even collection of traits is sufficient to make someone a misbeliever. The funnel is built of multiple elements, and personality is only one part of the story, albeit an important one.

Some of the Challenges of Experimenting with Misbelievers

One more word of caution on personality research and misbelief. Typically, if you were a social scientist trying to understand the role of personality in driving a certain behavior, you'd conduct research using detailed personality scales. As I have been writing this book, I have been trying to carry out this kind of research on the personality of Covid-19 misbelievers. However, it has turned out to be incredibly difficult. Maybe this should have been obvious to me from the get-go, but getting people who don't trust "the establishment" in any form to participate, even in a simple study, is, well, far from simple.

I began with a few of the people I affectionately thought of as my spirit guides in this unfamiliar realm—the misbelievers who talk with me and help me better understand their world. I asked them if they would be willing to help me with some research. At first they were skeptical but open to learning the details of the studies. I emailed them a link to an online survey that started by showing a long list of conspiracy theories (see page 178) and asking them to rate their level of belief in each. This was designed to measure general belief in conspiracy theories. Next the participants were asked to respond to at least one personality scale that measured one of many different personality traits, from loneliness to anxiety to social standing to trust, optimism, Machiavellianism, narcissism, and more.

My hope was to measure each participant's strength of belief in conspiracy theories and their score on some personality scales and then compute which personality traits were linked more strongly and weakly to misbelief. For example, would optimism be linked with high or low belief in conspiracy theories? Or would it be connected at all? What about trust? Social standing? Intellectual humility? Critical thinking? And so on?

The List of Conspiracy Theories I Included in My General Scale of Conspiratorial Beliefs

If you want to fill out this scale yourself, please write down your degree of belief next to each of the statements listed below using a scale from 0 (definitely not true) to 100 (definitely true). At the end, take the average of all the numbers, and you will have your level of conspiratorial beliefs score.

- Covid-19 was created as a bioweapon in a Chinese or US laboratory. _____

- Donald Trump really won the 2020 presidential election. _____

- The US moon landing was staged by NASA in a film studio. _____

- People are dying from other illnesses, but the deaths are being categorized as Covid-19 deaths. _____

- Covid-19 is a fake virus. _____

- Covid-19 is not as serious as it's being made out to be. _____

- The government covers up proof of alien life. _____

- Government agents threaten or assassinate witnesses who have seen proof of UFOs/alien life. _____

- Covid-19 vaccine side effects are being hidden. _____

- The science around global warming is invented or distorted for ideological or financial reasons. _____

- A group of international elites controls governments, media, and industry with the goal of creating global hegemony. These organizations include the World Bank, the United Nations, the European Union, and the Federal Reserve System. _____

- Mass shootings and other large tragic events are often staged, and victims and their families are really crisis actors. _____

- The 9/11 attacks were controlled demolitions of the World Trade Center. _____

- Several famous people's deaths are actually hoaxes, e.g., Princess Diana, Martin Luther King, Jr., the Notorious B.I.G., Kurt Cobain, Mozart, John Lennon. _____

- Fossil fuel companies are suppressing the progress of electric cars. Also, government agencies, special interest groups, or fraudulent inventors are suppressing perpetual motion and cold fusion technology. _____

- Radio frequency identification (RFID) chips, such as those people implant in their pets, are also secretly widely implanted in humans. _____

- The Illuminati was a secret society that formed during the Enlightenment, and they were responsible for the French Revolution. _____

- Dates and facts of historical events have been deliberately distorted. _____

- Under secret government policies, the water condensation trails from aircrafts are chemical or biological agents or contain a toxic mix of aluminum, strontium, and barium. _____

- The world is controlled by a secret cabal of Satan-worshipping pedophiles. Donald Trump is battling to stop this cabal. _____

- Immigration, integration, low fertility rates, and abortion are being promoted in historically white-majority countries to make white people a minority. _____

- Wealthy OECD nations, with only a small percentage of the global population, account for most of global GDP, and the richest people are responsible for more than half the world's carbon emissions. _____

- The United Nations will establish a new world order under its Agenda 21/2030 Agenda for Sustainable Development goals. _____

Once my misbelieving spirit guides saw the list, they wrote me back in alarm. What was I trying to study? Was I planning to make fun of them? How could they know that the research was going to be carried out in the right way? Richard was the person with the highest influence among them, so I scheduled a time to talk with him first.

"You do realize that you are trying to study people who don't trust the establishment, don't trust you, and don't trust much in general?" he asked.

"I do realize that," I replied, "and this is the reason that I approached you. I am hoping that if you asked your followers to participate, they would trust you."

"But what are you trying to study?" he demanded, moving to his next topic.

"I am trying to understand the personality structure in general and how it relates to people who have low trust in the government and society. I want to see which personality traits are stronger and weaker among such people." I tried to explain. (The truth is that I did want to study the links between personality traits and belief in conspiracy theories, but I was trying to frame the study in the most positive way I could.)

"It seems to me that you are trying to make fun of us," Richard protested suspiciously.

"Not at all. I just want to know which personality traits are more or less pronounced among individuals with low trust. There is no judgment, and these personality traits are not necessarily bad ones, they are just personality traits. I promise you that I am not here to pass any judgment, just to measure. Aren't you curious about what makes this group different from everyone else? Don't you want to know? Here is a chance to do research on this."

There it was—the magic word: *research*. That was my secret weapon to motivate Richard, and I was very pleased with myself. After all his talk about "doing research," I was sure he would not turn down the chance to do real research with a real researcher.

"But how could I trust the results?" he asked.

"Why don't you join me in this research and we can do it together? If we end up publishing it, you can also be one of the authors of the paper. And to be sure that you can trust the results, I will give you access to the online database so that you can verify all the data. What do you say?"

At that point Richard shifted his approach. Helping me with this research, he said, would tarnish his reputation within the group in a way that he was not willing to risk. That was a line of reasoning he had used before, about a year into our relationship, when I had asked him if he would help me clear my name within his group, since he had admitted to me that he no longer thought that I was part of the cabal. At the time he had responded somewhat apologetically, saying that he couldn't risk his standing in his community and that defending me would come with a social cost that was too high for him to pay. Now he was using the same logic in his refusal to help with the research, this time without the apologetic tone.

My discussions with other misbelievers I knew did not yield any better results, so I was left to try other methods. My desire to understand the relationship between different types of personality scales and the general belief in conspiracy theories was still strong. And so, armed with unrealistic hopes and a good dose of overconfidence, I went into this project together with Nina Bartmann and Shaye-Ann McDonald from my lab at Duke, the Center for Advanced Hindsight (as will become apparent very soon, being informed, as I am, about motivated reasoning and overconfidence does not provide a perfect cure for these biases). We started with David Icke's website. As a reminder, David Icke (who, interestingly, shares a birthday with me) used to be a professional athlete and sports broadcaster; at some point he started believing that there is an interdimensional race of reptilians, the Archons, that are controlling Earth. According to Icke, the Archons created a genetically modified human-Archon hybrid race of reptilian shape-shifters. He claims that these hybrids occupy positions of power in many professions in order to control the world and create a new world order. The good news about Icke's website is that it is possible to buy ads that will be shown to the people who visit, which we did. After spending a few thousand dollars on ads, we got six (yes, six, this is not a typo) participants to take our survey. We had a similar experience on Reddit, this time with eight participants, and Truth Social, the Donald Trump social network, gave us seven participants. Clearly, my confidence was misplaced and expensive.

These difficulties with conducting research on misbelievers are, of course, not my problem alone. Almost all the research on conspiracy theorists and misbelievers runs into this issue. As a result, much of the research focuses on mild to moderate misbelievers and not on extreme misbelievers. Why is this a problem? Consider the following illustration (see Figure 7, on page 183). On the horizontal axis is the strength of conspiratorial beliefs, ranging from 0 to 100. On the vertical axis is a certain personality trait, let's say the tendency to trust one's intuitions, also ranging from 0 to 100.

To properly carry out this research, we would ideally want people who score all along the horizontal axis: some who have a very low level of conspiratorial beliefs, some who fall somewhere in the middle, and some who score very high on their conspiratorial beliefs. If we had such a range, maybe we would get dots as depicted as panel A, revealing a very clear and strong relationship between the tendency to believe in one's intuitions and conspiratorial beliefs.

FIGURE 7

The problem with a biased sample

An illustration of the challenges that come with studying a relationship between two variables when people who hold the most extreme opinions don't participate in the study. Panel A represents the full range of conspiratorial beliefs and the resulting correlation. The shaded rectangle in panel B represents what the results would look like if they were based only on participants who held only low and medium conspiratorial beliefs. This is what we get when the extremes don't participate.

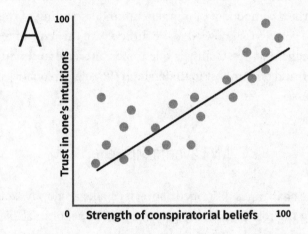

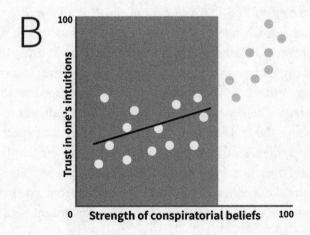

But what if the researchers had the same challenges as I did when I approached the misbelievers? What if, in the end, the participants in the study were only those with either low or moderate conspiratorial beliefs, because no one with strong conspiratorial beliefs participated? In such a case, we would see only part of the data (the shaded rectangle in panel B). In such a case, there would be a truncated range on the horizontal axis and therefore a truncated range on the vertical axis, and we would be unable to fully understand the relationship between belief in one's intuitions and conspiratorial beliefs.

Despite these challenges, understanding personality is certainly an important component of our general understanding of misbelief. So we are going to acknowledge the difficulty of this type of research, acknowledge that personality is one important force in the funnel of misbelief, and do our best to understand the role personality plays in the process.

My Favorite Misbelief

Given the challenges described above, let's take a sideways route into exploring the role personality plays in the funnel of misbelief. This path also gives me a chance to describe my favorite misbelief, which I'll introduce by asking you to imagine the following scenario.

It's the middle of the night. Suddenly, you awaken from sleep and you're alarmed to find that you cannot move at all. Your limbs are frozen, and a crushing pressure bears down on your chest, making it hard for you to breathe. You want to cry out, but your mouth won't open, nor can you make a sound. Despite this paralysis, you still feel sensations. Your whole body is tingling, as if it's plugged into an electrical current. The sensation is so powerful that it almost seems to lift you above the bed. Around you, where you might expect to see the familiar shapes of your bedroom furniture, there are flashing lights. Am I dreaming? you wonder. But your wildly pounding heart feels very awake, and your wide-open eyes swivel from side to side as you try to make sense of where you are. Try as you might, you cannot snap yourself out of the terrifying experience. A loud buzzing sound fills the room. Most frightening of all, you see strangely misshapen, shadowy figures gathered around your bed. They are intently focused on your body, and you are powerless to stop them as they probe you with unfamiliar instruments. Just as you feel that you can't bear the sensations and the panic any longer, the figures vanish, the lights are gone, and you feel the weight of your body sinking into the familiar softness of your bed. You regain the ability to move your hands and feet, and when you leap up to turn on the light, everything looks exactly as it should.

If something like this were to happen to you, what would you believe? Perhaps you'd decide that it must have been a dream. Or perhaps you'd come to a different conclusion: that you had been abducted by aliens. You may laugh. But it turns out that a surprising number of people believe just that. What is particularly odd about this misbelief is that all of the people who believe they were abducted by aliens tell relatively similar stories that go pretty much like the one described above. Consistent details include an inability to move, electrical tingling sensations throughout the body, feelings of levitation, loud buzzing sounds, flashing lights, and, most strikingly, seeing figures of aliens hovering near their bed and examining their bodies. People who have these experiences often believe that they were temporarily teleported into an alien craft and subjected to invasive medical experiments before being returned to their homes.

At first glance, it is rather tempting to discount these people and call them crazy. But there are lots of them, they describe very similar experiences, and they test within a normal range when they are evaluated for sanity. Finally, and perhaps most important, calling them crazy is in no way going to help us understand what is going on. When we do so, we lose the opportunity to learn from them. In many cases, studying extreme experiences is an opportunity to shed light on day-to-day experiences. For example, when we study marathon runners and mountain climbers, we can learn a lot from their motivations about the motivations of regular mortals who don't have such an intimate relationship with pain and frostbite. I've experienced this in my personal life as well. The extreme situations I found myself in when I was hospitalized after my accident have functioned as a magnifying glass to help me look at things such as pain, learned helplessness, and fear in ways that are applicable to regular day-to-day life.

Back to the alien abductees. Can we learn something from their experience that would apply to misbelief in general and to the relationship between misbelief and personality in particular? The answer is yes.

A group of researchers led by Susan Clancy examined the general experience described by alien abductees. (When we met, Susan told me that they wanted her to refer to them as "alien abductees" and not as "people who believed that they were abducted by aliens.") The researchers noticed that their descriptions of their experience (electrical tingling, feelings of levitation, loud buzzing sounds, flashing lights, hovering figures of aliens) seemed very similar to the descriptions of the known physiological situation called **sleep paralysis**.

What is sleep paralysis? During the periods of sleep when we dream (known as rapid eye movement, or REM, sleep), our brain is active and sends commands to the body to do different things: *Walk forward. Bend down to pick a flower. Draw your sword. Jump over the car. Punch the bad guy. Fly!* So how come we don't usually act out the commands from the brain during REM sleep? Why don't we find ourselves running around the bedroom or trying to launch ourselves out of the window? Because luckily, the human organism has adapted to prevent such dangerous situations. During REM sleep the brain essentially becomes disconnected from the body, which means that the body does not respond to the brain's signals.

Sometimes, however, things don't work as planned and the brain wakes up from REM sleep while the mechanism that paralyzes the body is still active. In such cases, for a very short time, the person is awake, to some degree, but cannot move. This is when people experience electrical tingling, feelings of levitation, loud buzzing sounds, flashing lights, and hallucinations. Interestingly, at earlier periods in history and in other cultures, hallucinations during this experience took different forms: people reported demons, witches, or ghosts sitting on their chests and preventing them from moving. Clearly, people describe images relevant to their time and culture, which are reflected in their interpretations of their experiences.

You might expect the researchers to declare victory here. After all, sleep paralysis and the reports of people abducted by aliens are incredibly similar, so maybe people who experience sleep paralysis misinterpret it as alien abduction—case closed. Not so fast. Why? Because according to the Sleep Foundation, at least 8 percent of the population experiences sleep paralysis at some point (and by the way, stress increases the odds of sleep paralysis), but the vast majority of these individuals never claim that they were abducted by aliens. How come? What separates those who interpret their experience as an alien abduction story from those who just get up and go about their lives with the same beliefs about the world that they had when they went to bed the night before? The answer to this question is where things get interesting and where we connect back to the relationship between personality and misbelief.

Susan and her group of researchers examined a few key personality differences that could potentially play a role in bringing about a conviction that a person had been abducted by aliens. They started with two ads: one that asked people to sign up for a memory study (the control group) and one that asked people who might have been contacted or abducted by aliens to participate in a memory study. That created two groups: the control group and the alien abductees (the original paper has a further distinction of different types of alien abductees, but this is not important for our purpose).

Once the two conditions were set, they were ready to study some of the individual differences. First, they examined certain aspects of faulty memory. Of course, all human beings experience imperfect memory from time to time, but particular tendencies toward misremembering are stronger in some individuals than others. In other words, some aspects of faulty memory can be considered a personality trait. The researchers wanted to know whether these memory-related traits were more pronounced in people who claimed to have been abducted by aliens.

Imagine that you are a participant in this experiment (I don't know if you believe that you've been abducted by aliens or not, so for now, let's just say you are a participant). As you enter the lab, you are given a set of instructions. For your first task, you will hear twenty lists of thematically connected words, each list containing between three and fifteen words, which will be played to you over a loudspeaker one at a time with a three-second interval after each word. You're asked to pay close attention, since you will be asked to remember the words. Once each list has been played for you, you are asked to solve four simple math problems. For example, how much is 16 + 52? After you finish solving the math problems, you are given ninety seconds to write down in your booklet all the words you remember from the list. Then the next list is played for you, followed by the next set of four simple math problems, and then you are asked to write the words you remember from that list, and so on for all twenty lists.

For example, here is one of the lists:

Sour, candy, sugar, bitter, good, taste, tooth, nice, honey, soda, chocolate, heart, cake, tart, pie

After being presented with all twenty lists, you have been exposed to 180 words in total. You might think that the main measure of interest is the number of words that you remembered correctly. This is indeed interesting, as a measure of your memory ability, but the research was not about remembering, it was about misremembering. The question was about the tricks that memory plays on us and the possibility that it can play more tricks on some of us than on others. The specific measure was the number of words that participants misremembered—words that were *not* presented on the list but that nevertheless participants remembered as if they had been presented. For example, if you wrote down "syrup" or "cookie" after hearing the list above, those would be misremembered words, since they did not appear on the list. Officially, this measure is called **false recall**.

False recall was not the only type of memory mistake that interested the researchers. They were also interested in **false recognition**. Here's how you would be tested for that. After going through the process described above with the twenty lists, you're given a master list that includes eighty words in total. Forty of the words are ones you have seen before, and forty are words that were not on any of the lists. As you look at these eighty words one by one, you're asked to indicate which were on one of the lists and which words are new. At the end of that exercise, the researchers had another measure of faulty memory: false recognition, represented by the number of words that you had not seen before but, when presented with them, you thought you had seen them.

In short, there are two different but related measures of faulty memory, both of them quantifying the frequency with which our minds wrongly confuse words we have not been exposed to with words we have been exposed to. The first, false recall, is based only on faulty memory search, and the second, false recognition, is based on reacting to words that are written on a page (no memory search is needed).

So what were the results? The rates of both false recall and false recognition were much higher for those in the alien abductee group. False recall was about twice as high and false recognition was about 50 percent higher in the alien abductee group. Together, these results showed that alien abductees recalled more fragments that were *not* in the original experience but nevertheless penetrated their memory. Put differently, they had a harder time separating things they had experienced from things they had not experienced. This particular piece of the puzzle might help explain how somebody who wakes up in a state of sleep paralysis with all kinds of images lingering from their dreams plus the strange sensation itself might create a picture of their experience that includes false elements, while feeling as if it were part of their real memory and real experience.

This might seem like a good place to close down the alien abductees investigation and celebrate explaining the mystery. But the researchers did not stop with memory. They continued exploring other elements of personality as well, which is why their research is helpful in our exploration of the broader phenomenon of misbelief. But before we get to that research, let's follow the theme of misremembering a little further.

The Real-Life Dangers of Misremembering

How might confused memories like the ones described above play a role in daily life? And how could such false memories play a role in ushering people down the funnel of misbelief? Good of you to ask. Let me tell you a story about a conversation I had with a woman named Jaime.

Jaime was the leader of a very influential group of misbelievers that was supposedly collecting testimonies from people who claimed to have experienced Covid-19 vaccine side effects. She told me she had collected 5,500 testimonies of terrible side effects ranging from muscular challenges to long-term illness to death, all associated with the Covid-19 vaccine. Not only was she collecting these testimonies, she was also echoing them and distributing them anywhere she could. In my mind, she was leading people to believe that the side effects were much more severe and more common than they really were. I also thought that in all likelihood, her reporting included all kinds of side effects from life, in general, and from living through the pandemic (increased comfort eating, more loneliness, higher stress, inactivity, and so on) that were not in fact directly caused by the vaccine. In talking to Jaime, however, my intention was not to challenge her activities and point out the damage she was causing, but rather to try to understand how she had gotten into that line of work. What process had led her to misbelieve to such an extent that she spent a large part of her time documenting side effects and distributing the information? Pre-Covid, she had been working at a high-tech company analyzing marketing trends. By all accounts, she was very much a member of the establishment. And now, just two years later, she was one of the most extreme spreaders of fake news regarding the Covid-19 vaccine and its side effects.

"How did you get here? What were the first steps you took?" I asked her.

I also asked her to recall and try to describe the first "red flags" that had made her worry that things in the world were not as they seemed. Early in the pandemic, she told me, like everybody else, she had been waiting for a vaccine to be approved and save us all. She had expected it to take years, so she was very excited that the process happened so fast. As soon as a vaccine was approved, she went onto the FDA website to read about it. What she read worried her. She noticed that the FDA clinical trial of the vaccine had some shortcomings in terms of how it was conducted. She was particularly worried about the control group that had been abandoned and not reported (a claim for which I could not find supporting evidence). In that moment, she told me, she decided that the vaccine was not the vaccine she'd been hoping for.

I probed her memory a little: "When exactly did you read all this information on the FDA website? And had you ever read any information from the FDA before?"

"I read it the day that the FDA announced the vaccine was approved," she replied confidently. "It was all there in their documents." Of course I asked for the links, but she never sent them to me.

"And what happened next?" I asked.

"A few days later, I listened to a lecture by a very famous doctor, somebody who is an expert on these matters and who knows a lot about vaccines and their side effects. In her talk, that doctor took apart the Pfizer methodology and described all the ways in which the clinical trial was faulty and designed to get the expected result without really testing it."

"Did this doctor also mention the abandoned control group?" I asked.

"Yes, she pointed out that there was no real control group and the one control group that was created correctly was dropped from the analysis."

Then I pressed her on her memory: "Did you actually read about the abandoned control group in the FDA documents, or was it this doctor who told you about all the faults and the missing control group?"

Jaime looked confused. "I don't remember. I think maybe it was both."

This is the kind of memory confusion we saw earlier, where something that didn't happen (reading about an abandoned control group in FDA documents related to the approval of the Pfizer vaccine) is confused with something that happened (hearing about possible faults with the research from someone). My guess is that Jaime never

---

HOPEFULLY HELPFUL

Risk Assess Your Friends

Because personality is difficult to change, it's important to be aware of the personalities of those around us and pay particular attention to those who are most susceptible to going down the funnel of misbelief. (We'll look at more examples of what traits make people susceptible in the next chapter.) The recommendation here is not to avoid friendships with people who display these traits; after all, the personality characteristics that make people susceptible can be quite positive. Some, such as narcissism or a tendency to misremember, are largely negative. But others, such as creative ways of thinking and connecting dots or even a tendency to be suspicious, can in many cases be useful and even desirable. It's only when other factors, such as stress, begin ushering a person down the funnel that these traits begin to operate in more negative ways.

So if you notice that a person you care about seems to have personality characteristics that might render them at a high risk for going down the funnel of misbelief, pay extra attention to this person and try especially hard to prevent them from going down the funnel. The earlier you can intervene, the better.

read any of the original reports on the FDA site, even though her memories of reading those documents and catching the problems are as vivid as if she had perused them cover to cover. She heard the doctor making some claims and created a false memory; the same type of misremembering we saw with the alien abductees.

What makes this a serious issue is that the tendencies to misremember and to trust one's memory unquestioningly are not limited to recalling some words from a list. As we see with the alien abductees, it can extend to events that may cause serious damage to a person's life or reputation. As was evident with Jaime, it can extend to important facts relating to the effectiveness and side effects of a vaccine. And because Jaime became one of the most respected figures within the community of misbelievers, a hub for collecting and spreading information for many people, the effects of her faulty memory were substantially greater, with devastating consequences for herself and for the people who listened to her.

The Personality of Alien Abductees: A Deeper Dive

We've established that those whose personalities are prone to misremembering may be more susceptible to misbelief. But what about other personality traits? The researchers studying the alien abductees asked participants to fill out many additional different personality scales. Their aim was to figure out what kinds of personalities are pronounced among those who believe that they have been abducted by aliens. From this exercise, three main personality traits stood out: **magical ideation, openness to absorbing**, and **perceptual aberration**. In general, these personality traits measure the extent to which people believe in unconventional forms of causation; are more easily absorbed in their mental imagery and fantasy; are more easily hypnotizable; and believe that certain people have special powers.

Here are some examples from these types of personality assessment tools. You can try them for yourself. For each of the questions, rate yourself on a scale of 0 (does not describe me at all) to 100 (describes me very well). The higher the average of your answers, the higher your score on these general personality traits.

- Some people can make me aware of them just by thinking about me. _____

- I have sometimes been fearful of stepping on sidewalk cracks. _____

- Horoscopes are right too often for it to be a coincidence. _____

- Things sometimes seem to be in different places when I get home, even though no one has been there. _____

- Numbers such as 13 and 7 have special powers. _____

- I have felt that there were messages for me in the way things were arranged, as in a store window. _____

- Good-luck charms work. _____

- It is possible to harm others merely by thinking bad thoughts about them. _____

- I have sometimes sensed an evil presence around me, although I could not see it. _____

- I sometimes have a feeling of gaining or losing energy when certain people look at me or touch me. _____

- At times I perform certain little rituals to ward off negative influences. _____

- I have felt that I might cause something to happen just by thinking too much about it. _____

- My average score across these questions is _____

So what does all this research tell us? Taken together, it suggests an interesting mechanism that starts with an anchor in reality (the real experience of sleep paralysis) but takes off from there, assisted by the stress that some people feel when they experience sleep paralysis combined with a few personality traits that take this complex experience and convert it into a memory of alien abduction.

The findings from the study of alien abductees are incredibly important because they provide a clear mechanism for something that previously looked like pure nonsense. Did this research convince all those alien abductees that little green men had never showed up in their bedrooms? No. People are attached to their misbeliefs, and it takes more than a rational explanation backed up by a study to disabuse them of their stories. But for those of us who are interested in understanding the phenomenon of misbelief, this study is enlightening. And the story does not end here. Over the next several years, more and more researchers examined all kinds of personality traits that are related to all kinds of beliefs and misbeliefs. These will be the topic of our next chapter.

CHAPTER 8

An Attempt to Classify the Role of Personality in the Funnel of Misbelief

Human beings are pattern-seeking animals who will prefer even a bad theory or a conspiracy theory to no theory at all.

—CHRISTOPHER HITCHENS

Have you ever lain in a field on a summer afternoon looking aimlessly at the clouds floating in the sky above? My guess is that after a few minutes of this, your mind started searching for a causal story that described what was going on above your head. "This little cloud is so cute and happy, but why is the large cloud chasing it? Run! Run! Why aren't the other clouds helping this cute little cloud? Don't they care? Oh, maybe this other cloud headed in this direction is a friend and he is coming to help him out. Never mind, the little cloud is clever and he managed to escape."

If you've ever played out scenarios like this, don't worry, you're not alone! What you've experienced is the very common human tendency to see patterns where none exist—a characteristic that the author and *Skeptic* magazine publisher Michael Shermer calls **patternicity**. As in the case of misremembering, which we examined in the last chapter, patternicity is both a general characteristic of human nature and a personality difference, in the sense that some people display a much greater propensity to see patterns than others, especially when they are under stress. Before we look more closely at patternicity as a personality trait, let's familiarize ourselves with it as a general human characteristic.

Perhaps you're not a lying-on-the-grass-gazing-at-the-clouds kind of person and you didn't relate to my example. Instead, consider the following study that I carried out with some MBA students. The students were invited to the lab to play a stock market game. We asked them to imagine that the stock market had just closed, and then we showed them the behavior of one stock throughout the day. In some cases, the stock price went up; in other cases, it went down. After showing them the behavior of one such stock, we shared with them an analysis of the stock by a confident, Jim Cramer type of analyst. And here comes the important part of the study: for half of the participants, the analyst gave reason X for why the stock price had done what it had, and for the other half, he gave reason not-X, which was the opposite of X. For example, he might have said, "IBM stock went up because IBM just *bought* back some of its own stock and the market reacted positively to this news." For the other participants, the analyst gave the opposite reason for the exact same behavior of the stock price. For example, he might have said, "IBM stock went up because IBM just *sold off* some of its own stock and the market reacted positively to this news."

The process was repeated with many participants, using lots of stocks and lots of after-the-fact stories for the behaviors of the stock prices. Then the MBA participants rated how much sense the reasons made to them. And guess what? They were equally highly impressed with the logic of each reason and with that of its opposite. Of course, it is easy to make fun of the stock market pundits who explain things with unshakable confidence just after the market closes, but that was not the point. The point is that our minds are always looking for stories. This is true for all of us, including the pundits. The moment that a story starts to build up (IBM bought back some of its shares), we don't think critically but instead keep on looking for other elements that could fit with the story (more evidence that our little cloud is indeed a cute one and the larger one is evil). Our mind is driven to try to find stories everywhere we go. To be clear, the fact that our mind works this way has some very important benefits. We are wired to find connections, and to a large degree this is how we learn about the world: we see a pattern, we suspect a causal relationship, and sometimes we discover new things.

This is how science and technology move us forward. Someone notices a link between a petri dish that was left alone for a while and yeast, and soon enough we have antibiotics. Someone realizes that mechlorethamine, a chemical warfare agent, works by binding to DNA, cross-linking two strands and preventing cell duplication, and now we have a chemotherapy drug used to fight lung disease, Hodgkin's disease, and non-Hodgkin's lymphoma. Researchers create sildenafil citrate, which they believe will be useful in treating high blood pressure and angina, a chest pain associated with coronary heart disease, they pay attention to the side effects, and we get Viagra. And when wine growers from one part of France try to make wine equal to the quality of Burgundy wines but fail because the colder winters stop the fermentation prematurely, they notice that something interesting is happening and make a new wine, which they name after their region, Champagne.

What should also be obvious from the stock market experiments is that this ability to see patterns and connect dots comes with a substantial price. The price is that sometimes we find connections that simply do not exist. And sometimes we are so sure that they exist that we keep insisting on the links, despite all available evidence to the contrary.

Now that we've established the human propensity to see patterns, what about individual differences? What makes some people more likely than others to see patterns, especially where none exist? And what about the link between conspiratorial thinking and observing illusory patterns? This was exactly what Jan-Willem van Prooijen, Karen Douglas, and Clara De Inocencio set out to test.

In one of their experiments, the researchers presented participants with a random sequence of one hundred coin tosses, represented by *H* for "heads" and *T* for "tails" (for example, HTHTHHTTHHHT-THTHHTTT . . . T). Then they asked the participants to indicate to what degree they thought the sequence was random or predetermined. The result gave each person a score for seeing patterns where none exist, or in short, patternicity.

They also asked the participants to respond to many questions, among them a set of invented "facts" about the energy drink Red Bull (they made up facts about Red Bull to minimize the chance that their participants had any prior beliefs about the statements). For each statement, they asked the participants to report the degree to which they thought the statement was true.

Try it for yourself: For each statement below, please indicate the degree to which you believe it is likely to be true or false using a scale from 0 (definitely not true) to 100 (definitely true).

- Red Bull contains illegal substances that raise the desire for the product. _____

- The official inventor of Red Bull pays 10 million euros each year to keep food controllers quiet. _____

- If a can of Red Bull is heated up to 104°F, it releases health-threatening substances. _____

- Subliminal messages in Red Bull television commercials make consumers believe that Red Bull improves one's health. _____

- The slogan "Red Bull gives you wings" is used because in animal experiments, rats grew rudimentary wings. _____

- Regular consumption of Red Bull raises dopamine levels, which causes damage in the long term. _____

- In the beginning, Red Bull was illegal for minors in America, which raises questions as to its subsequent legalization. _____

- Commercials in sports give the impression that Red Bull is healthy. _____

- The extract "testiculus taurus" found in Red Bull has unknown side effects. _____

- My average score across these questions. _____

The researchers then created a "Red Bull conspiracy score" for each participant by averaging their responses to the Red Bull conspiracy questions and then examined the correlation between their Red Bull conspiracy score and their patternicity score. As you might have expected (sometimes when our mind keeps looking for patterns, it leads us in the right direction), they found a very substantial correlation. This shows that the less trusting mind, represented by those who were more suspicious of Red Bull (and presumably also less trusting of anyone and anything), is also more likely to see patterns where none exist, such as in a random sequence of coin tosses.

More generally, these results suggest that minds that are more likely to see patterns where none exist are also minds that are more suspicious. But before we start feeling bad for people saddled with these personality traits, let's be clear: seeing patterns where none exists (patternicity) and being more suspicious are in and of themselves neither positive nor negative. It is only the decisions that people make based on these perceptions that could have positive or negative impacts. And sometimes it is not immediately obvious what is good and what is bad.

Consider my great-grandmother Mina. She was a very suspicious person who saw patterns all around her. She was so sure that terrible things were about to take place that she insisted that her immediate family pack up their lives and move to another country—twice! Good or bad? Well, here are some more details:

My great-grandmother was born in Russia in 1890. As the Russian Revolution of 1917 started, she terrorized my great-grandfather into leaving Russia, which they did. Where did they escape to? Germany, of all places. Some years later, as Hitler was rising to power, Mina got into such a state of fear that she convinced her family to pack everything they owned that day and leave Germany overnight. For a second time, she very likely saved everyone around her. Her heightened propensity to see patterns and be suspicious was truly a blessing. Though, to be fair, the story about my beloved great-grandmother is more complex and not everything about her patternicity was a blessing. Regardless, you can see that these personality traits can have positive or negative outcomes, depending on how they are used and for what purpose. I am personally very grateful for the patternicity and suspicion of my great-grandmother and at the same time very upset about the people who accuse me of perpetrating a Covid hoax.

A final note on being suspicious and patternicity: it can be hard to tell which one causes the other. Is it the case that people who are more suspicious keep looking for potential threats and as a result they see patterns in all kinds of places (i.e., suspicion → patternicity)? Or is it the case that people who see patterns around them, especially patterns that other people don't see, become more suspicious as a result (i.e., patternicity → suspicion)? It's hard to say, and most likely it goes both ways. What is clear is that the two traits work together to push people down the funnel of misbelief.

How Stress Increases Patternicity

We've established that the unholy alliance of patternicity and suspicion plays an important role in accelerating the journey down the funnel of misbelief. Importantly, there are other forces that can add to the potency of patternicity and suspicion and make them even more powerful and dangerous. What might such forces be? If you think back to the chapters on stress, you will probably predict that one of the most important of these forces is lack of control, which, as it turns out, also leads to increased patternicity.

To look at the possible links between lack of control and patternicity, Jennifer Whitson and Adam Galinsky carried out a few fascinating experiments. They posited that when people are *objectively* out of control because of factors in their environment, they will try to regain even a perceptual sense of control. And they will do so by seeing patterns, including seeing images in random dots, perceiving conspiracies, and adopting superstitions. Before describing the experimental evidence, their paper lists some fascinating natural observations that support the general links between lack of control and patternicity. For example, in a classic text on science and religion, Bronislaw Malinowski described the tribes of the Trobriand Islands and their fishing rituals. These tribes fish in two main environments: deep sea and shallow waters. As you might expect, deep-sea fishing is much more unpredictable and depends to a larger degree on weather conditions, therefore giving the fishermen a greater sense of being out of control. Fishing in shallow waters is relatively more predictable and consistent, therefore making the fishermen feel more in control. And guess what kind of fishing has a larger number of rituals associated with it? You guessed it: deep-sea fishing. Presumably the greater unpredictability creates the need for a feeling of control, so the fishermen turn to superstitious rituals to counteract the lack of control.

Another slightly more amusing example came from Pavel Simonov and his colleagues. They studied an activity that most of us would associate with being highly out of control: parachuting. Consistent with what we know already, they found that parachute jumpers are more likely to see figures in images that are made only of random dots. Providing further support to this link, they also found that they see even more of these figures just before stepping out of a plane— clearly a stressful moment without much control.

When it comes to superstitious rituals, few activities involve more than the beloved sport of baseball. There are many well-known superstitious rituals in the baseball world: refusing to wash uniforms during a winning streak; never stepping on the foul lines coming onto or going off the field; abstaining from sex on game day; hitters drawing symbols with their bat prior to batting; using lucky bats and gloves; and so on. What is particularly interesting is an observation by George Gmelch that, much like the Trobriand Islands fishermen, players who play in positions that are more unpredictable (for example, the pitcher and batters) have a higher number of these superstitious rituals, whereas players who play in positions that are more predictable (for example, outfielders) have fewer superstitions.

Though it is very difficult for lab experiments to be as much fun as these natural observations, Whitson and Galinsky did their best to make their study interesting. Here's how it worked: Imagine that you are taking part in an experiment. For the first task, you are told that the computer in front of you has chosen a concept. What is the concept? You don't know. Your task will be to guess what it is over the course of ten trials. This sounds a bit stressful, but you are also told that the only goal of the experiment is to understand your intuitive judgment, and with this additional information, you relax and press the start button.

For each of the ten trials, the computer shows you two shapes, one on the right side and one on the left side of the screen. Your task is to pick the one that you think belongs to the concept you're trying to identify. You pick, and at the end of the ten trials the computer picks the next concept and you repeat the whole thing for a total of five concepts and fifty trials.

This is the control, nonstressful condition. What about the experimental condition? In this case, the experiment is designed to give you a lower sense of control. If you are assigned to this condition, you are not given the relaxing statement about the researchers being interested in your intuitive judgment. Instead, you are told that there are correct and incorrect answers and that your task is to be correct. You are also told that the computer will tell you if you get the answer right or wrong, and this feedback is supposed to help you guess better next time. This already sounds more stressful, doesn't it? But it gets worse. What you don't know is that there actually is no concept, the patterns are selected at random, and so is the feedback you are given by the computer. Fifty percent of the time you are told that you are right and 50 percent of the time you are told that you are wrong, but none of it helps you guess the concept—because there is no concept! With the passing of each trial, you feel more and more unsuccessful, confused, and frustrated. Of course, it's not your fault, but you don't know that. Basically, the computer is gaslighting you.

Once this task is over, you move to the next task, which is a visual perception task and is the same for all participants. You are shown a set of grainy or "snowy" computerized pictures in which it is very difficult to make out a figure, if there is a figure at all. For this task you are asked to identify the figure in each picture and, if there isn't one, to indicate as much.

As you might expect, these two tasks were placed in the same experiment for a reason, and their order was crucial. The overall goal of the experiment was to examine how a diminished sense of control in the first task influenced the need to regain a sense of control by seeing patterns in the second task, even when none existed. Now, it must be said that the level of diminished sense of control in that experiment was rather modest compared with the actual levels of such feelings when they come bundled with real-life situations such as Covid-19, serious illness, job loss, and so on. Nevertheless, it is important to test whether even such a relatively mild sense of diminished control could create an impact. And did it matter? Of course it did! Our sense of control is so important to us that shaking it even a little has a deep impact. The participants with the lower level of perceived control were much more likely to see shapes in the snowy, grainy images. Having had the stressful experience of failing to figure things out during the concept task, they felt a need to regain the sense that they understood the world. So their minds worked extra hard, so hard that they invented figures where none existed.

Whitson and Galinsky carried out many other experiments to further explore this topic, including one showing that after recalling a personal experience involving a lower level of control, participants tended to become more superstitious. Overall, their results show that the mind is a sense-making organ that searches for patterns all the time, and therefore all of us see patterns. But when we feel a low level of control, our minds work even harder to find patterns to help us make sense of the world, even if it is made up.

Trusting and Overtrusting Our Intuitions

"From the beginning, I *knew* that something was wrong. I just felt it." So began one of Jenny's messages to her followers in July 2022, about two and a half years after the start of the Covid-19 pandemic. At that point, Jenny had shifted from trying to fight Covid restrictions and the vaccine to trying to establish her leadership role among the misbelievers. To that end, she placed great emphasis on the power of her intuitions. "I had a lot of prior experience with psychology and NLP [neuro-linguistic programming] and I could just feel that something was wrong," her post continued. "First with the restrictions and rules, then with the rag that they made people wear to block clean air and oxygen to their brain, and then of course with the experimental Covid-19 vaccine. I don't know what it was exactly, a kind of a feeling, but it was very strong and I knew no one should take such experimental potions. I trusted myself to follow what I felt and many of you did too. I know that some of you woke up after taking one jab, and some woke up after taking two or even three jabs, but we all know what we are feeling and we trust what we know and no one can tell us that what we know is not the truth."

Among the misbelievers, this type of internal conviction is common, coupled with a full disregard of external data. One of the fake health gurus I listened to while trying to understand their odd world (and the even odder world of their followers) asked his audience to realize that they didn't need to be an expert, a doctor, or a biologist to understand what he was telling them about the ways in which the body can heal itself and about the dangers of modern medications. The only thing they needed to do was to listen to themselves and trust their intuitive understanding. Another woman, who claimed to be an expert in "evolutionary astrology," gazed unblinkingly into her phone's camera as she explained to her audience that learning to trust their "inner guidance system" requires a traumatic separation from the consensus-believing herd, who find a sheeplike security in trusting the government, experts, celebrities, and mainstream media. It's not just New Age social media influencers who make such claims. Donald Trump often privileged his intuitions, which he also referred to as his "common sense," "hunches," or "gut instinct," over the data that were presented to him. As he memorably told the *Washington Post*, "I have a gut, and my gut tells me more sometimes than anybody else's brain can ever tell me." (To be clear, we all feel very confident in our gut intuition from time to time, even when it points in the opposite direction from the available data, but we don't always feel confident enough to say it out loud, and hopefully we don't unquestioningly act based on our gut intuition.) As I listened to more of these kinds of statements (an occupational hazard while doing the research for this book), one night I even dreamed that a student who got a C in my class came to see me about his low grade, telling me that inside he felt like an A student. I woke up (but not really).

Intellectual Humility

This tendency that Jenny, the health guru, the astrology practitioner, Donald Trump, and my imaginary student exhibit has been called many different scientific-sounding names. It used to be called *intuitive thinking*. The fashionable term that is currently in use is **intellectual humility** (or more accurately, in the cases mentioned, the lack thereof). Intellectual humility sounds much more politically correct while also sounding sort of judgmental. The basic idea behind the term is that people who possess a high level of intellectual humility recognize to a significant degree that their own beliefs and opinions might be incorrect.

Let's admit it: we all walk around the world with beliefs and opinions about the best way to invest our money, the correct party to vote for, the right way to raise kids, the superior phone, and the optimal diet. And as mentioned in chapter 6, our high level of confidence is often uncalled for (remember the Dunning-Kruger effect and the illusion of explanatory depth?). Those of us with a high level of intellectual humility (and who among us would not like to think of ourselves as high on intellectual humility?) walk through the world with maybe the same basic beliefs but with less conviction and more openness to changing these beliefs.

With this in mind, we can see how intellectual humility describes an overall mindset or a personality characteristic that is evident across many aspects of life and many judgments and decisions.

Here is a way to figure out your level of intellectual humility. Consider the following statements, based on the Comprehensive Intellectual Humility Scale, and for each of them, indicate truthfully the extent to which it describes you, using a scale from 0 (strongly disagree) to 100 (strongly agree). In this case, a high score = high intellectual humility (a rather contradictory term, I know).

- My ideas are usually not better than other people's ideas. _____

- For the most part, I have more to learn from others than others have to learn from me. _____

- Even when I am really confident about a belief, there is a chance that my belief is wrong. _____

- I'd rather turn to others for expertise than rely on my own knowledge about most topics. _____

- Even on important topics, I am likely to be swayed by the viewpoints of others. _____

- I have at times changed opinions that were important to me, when someone showed me that I was wrong. _____

- I am willing to change my position on important issues in the face of good reasons. _____

- I am willing to change my opinions on the basis of compelling reasons. _____

- I respect that there are ways of making important decisions that are different from the way I make decisions. _____

- Listening to perspectives of others often changes my important opinions. _____

- I welcome different ways of thinking about important topics. _____

- I can have great respect for someone, even when we don't see eye-to-eye on important topics. _____

- Even when I disagree with others, I can recognize that they have sound points. _____

- When someone disagrees with ideas that are important to me, it doesn't feel as though I'm being attacked. _____

- I don't tend to feel threatened when others disagree with me on topics that are close to my heart. _____

- I am willing to hear others out, even if I disagree with them. _____

- I feel just fine when others disagree with me on topics that are close to my heart. _____

- When I am given a scale on intellectual humility, and deep down I really want to find out that I am one of those humbler intellectuals, I am nevertheless able to resist the urge to inflate my answers (this is not a real part of the scale, just testing that you are paying attention). _____

- My average score across these 18 questions. _____

In general, people who score high on intellectual humility are more likely to believe that their beliefs might be incorrect; pay more attention to the strength of the evidence presented to them; be more interested in why people disagree with them; and devote more attention and time to views counter to their own. With all of these differences, it is easy to see how those with higher intellectual humility are more open to hearing others' opinions and more likely to reconsider their beliefs, opinions, and behaviors when they are given compelling reason to do so. After all, since they are inherently flexible about their preexisting opinions, they are less committed to these ideas and the door is open for them to be persuaded that they might have had the wrong ideas all along. Of course, as with most things, intellectual humility can be taken too far. In a society such as ours, which is plagued by overconfidence, it is unsurprising that we would benefit dramatically from a higher dose of intellectual humility. But we also don't want to end up in a world like the one William Butler Yeats described in "The Second Coming," in which "The best lack all conviction, while the worst / Are full of passionate intensity." With this in mind, we should aim for a balance between intellectual humility and conviction across the belief spectrum.

What about the relationship between intellectual humility and conspiratorial thinking? As you might expect, and as Shauna Bowes and her colleagues empirically showed, the strength of beliefs in conspiracy theories is negatively related to intellectual humility. That is, those who are higher in intellectual humility are less likely to believe in all kinds of conspiratorial thinking, fake news, misinformation, and pseudoscience, while those who are low in intellectual humility are more likely to believe in all of the above.

One complicating factor in understanding misbelievers is that they often speak in ways that appear to reflect intellectual humility. Often in my interactions with misbelievers, I found it interesting (and annoying and worrisome) to hear statements such as the following:

"I am not saying that masks don't reduce the spread of germs, I am just saying that we don't yet have enough data."

"I am not saying that the government planned 9/11, I'm just highlighting the fact that jet fuel can't melt steel."

"I am not saying that the vaccine is changing people's DNA, I just think we need to wait until we are sure."

"I am not saying the moon landing was definitely a hoax, but have you looked closely at those videos? Isn't it plausible that they *could* have been created in a studio?"

"You have a theory and I have one, why should we assume that yours is more accurate than mine?"

These, of course, are not statements that reflect authentic intellectual humility, but they pretend that they are. The people who say things like this are trying to appear open to other opinions and to give the impression that they are participating in an intellectually honest debate. They use the language of intellectual humility to camouflage the fact that they're not even interested in a discussion. If you find yourself in such an exchange, my advice is to cut your losses and leave, unless you are trying to understand the psychology of the other side, in which case give up on the topic of the discussion and just try to understand how the other party came to believe what they currently believe.

At this point, you are probably thinking to yourself that since you are high in intellectual humility, you are immune to misbelief. Not so. (It would actually reflect more intellectual humility to accept that we are all vulnerable to misbelief.) At the risk of repeating myself, I want to emphasize that as in the case of many other personality traits we've examined, although the correlations between low intellectual humility and misbelief are robust, they are also small to medium in size. This suggests that low intellectual humility is an important building block of susceptibility to conspiratorial thinking but certainly

---

HOPEFULLY HELPFUL
Practice Intellectual Humility

The old adage "Fake it till you make it" can be useful when it comes to increasing intellectual humility. Just inserting certain phrases into our conversations is a good start. Try these: "I'm not sure"; "I may be wrong"; "I wish I knew more about the topic"; "To the best of my knowledge." You may not even believe them at first. But saying them can serve as a reminder that perhaps we don't know as much as we think we do, and such statements can also shift the general tone of the conversation, inviting more intellectual humility in others as well.

Another way to increase intellectual humility is, from time to time, to take the opposite position from the one we are very sure about and argue against ourselves to the best of our ability. An even more potent form of this approach would be to make these arguments publicly in front of our friends, but if this is too embarrassing, try arguing in private to start with.

A study by Tenelle Porter and Karina Schumann tested another approach that entailed adopting a growth mindset attitude toward intelligence. In their study, participants who were encouraged to view intelligence as a psychological construct that can be improved over time (growth mindset), as opposed to viewing it as static and unable to be changed over time (fixed mindset), exhibited greater intellectual humility and were more open to learning from an opposing point of view.

not the whole picture. Even if someone is low in intellectual humility, it is too early to conclude that they are inevitably going to become a misbeliever. One more point about the research by Bowes and her colleagues: they also found that narcissism is also related to conspiratorial thinking, something we'll return to later in this chapter.

The Cognitive Reflection Test

One other interesting measure related to the general construct of intellectual humility is what Shane Frederick calls the Cognitive Reflection Test (CRT). The basic Cognitive Reflection Test includes three simple math questions. As an example, try to quickly answer this one:

A bat and a ball cost $1.10. The bat costs $1.00 more than the ball. How much does the ball cost?

Quick. What's your answer? _____

If you said "Ten cents," you're not alone. This is a common answer, but it's not correct. It's the first answer that comes to mind, but not the right one. With the numbers $1.10 and $1.00 running around in your head, your intuition prods you to answer "Ten cents." And some people stop at that intuitive answer. Other people check their answer. If you checked your answer, just to be sure, you most likely went through something like the following thought process: "It seems to me that $0.10 is the right answer, but let me check it first. If the ball were $0.10 and the bat is $1.00 more, then the bat would be $1.10, which, combined with the ball, would equal $1.20, not $1.10 (0.1 + (1 + .1) = 1.2)! This is not right. Together the bat and the ball have to be cheaper by $0.10, so maybe the ball is $0.05 and the bat is $1.05? Let's check this option (0.05 + (1 + 0.05) = 1.1). Yes, it is five cents and a dollar and five cents. Let me give this as my answer."

The point of the CRT is that it is the kind of math that everyone can do correctly if they want to. It is not a test of math ability; it is a test of the degree to which we trust our initial intuitions and allow them to guide our decisions, untested and unverified.

Here are the other two questions from the CRT (after you give your answers, check the footnote for the correct answers[*]):

If it takes 5 machines 5 minutes to make 5 widgets, how long would it take 100 machines to make 100 widgets?
Answer: _____

In a lake, there is a patch of lily pads. Every day, the patch doubles in size. If it takes 48 days for the patch to cover the entire lake, how long would it take for the patch to cover half of the lake?
Answer: _____

At this point, you might wonder about whether the CRT could predict belief in conspiratorial thinking. In terms of the logic, it seems to fit. Some people have a deep trust in their intuition (remember Jenny's statement earlier in this chapter: "We trust what we know and no one can tell us that what we know is not the truth."). Much like those who stick with the ten-cent answer to the first CRT question as soon as it jumps to mind, some people have such deep faith in their gut feelings and intuitions that they don't check their own answers. This general hypothesis was the exact thing that Gordon Pennycook and David Rand tested. As you might expect, they found that a lower score on the CRT (trusting intuitive answers more and rechecking less) is indeed connected with a higher level of conspiratorial beliefs.

---

[*]The intuitive answers to the second and third CRT questions that people typically give are 100 minutes and 24 days, while the correct solutions are 5 minutes and 47 days.

Decision-Making Skills

Success in life is based largely on taking in the information around us and correctly combining it in the right way in order to make good decisions. Indeed, a lot of life can be thought of as making good or bad decisions, and decision-making is a perspective from which we can consider almost anything we do. This is why this short section will look at a few topics we have already discussed but do so from a decision-making perspective. How does decision-making relate to personality? Well, it's an important individual difference. Some people are better at this skill than others.

It is easy to see how personality traits that capture the ability to make good decisions can play an important role in overall success in life: when to wake up in the morning; whether to hit the snooze button or not; how much time to spend brushing our teeth; whether to weigh ourselves or not; whether to take our medications or not; what to eat for breakfast. And then there are the large decisions, such as who to marry, what house to buy, whether to have kids and how many, how far to live from our mother-in-law, where to invest our retirement fund, and who to make the beneficiaries in our will.

When it comes to our decisions, it is rare for us to consider all the pros and cons, combine them in an optimal way, and arrive at a deliberate decision. Even in cases where we write down all the pros and cons, we often just pretend that we are taking all the elements into account. Most of the time, we make our decisions quickly, based on heuristics. What are heuristics? Heuristics are shortcuts. To make fully deliberate decisions about everything would take a long time and a lot of effort, so heuristics offer us a way to make quicker, generally okay decisions while feeling confident that we have made fantastic decisions. For example, let's say we go to a store and look at two bicycles. The spec sheet is very long and complex, and we are not sure we even know all the terms (cassette, derailleur, and Schrader, to mention just three). So how are we to choose? We pick the brand name we've heard more often. Or we go for the more expensive bike. In fifteen minutes, we have made a decision and we are out of the store, plus we feel good about our decision-making ability. These are heuristics—quick-and-dirty mechanisms that make it easier to make fast decisions while sacrificing some degree of decision quality. Of course, sometimes heuristics provide us with a quick decision without having to sacrifice too much in terms of our decision's quality (for example, when the brand of the bike is a really good indication of its value) and sometimes they get us into trouble (for example, when we assume that the most expensive option is the best one).

Heuristics are decision-making shortcuts that we all use, but they are also linked to personality traits in the sense that some people use them more than others, to both their benefit and their detriment. As a sample of the link among heuristics, personality traits, and a faster journey down the funnel of misbelief, consider the following three types of decision-making errors: the **conjunction fallacy**, **illusory correlations**, and the **hindsight bias**. This is not an exhaustive list but is intended to illustrate the more general point about decision-making and one's beliefs.

The Conjunction Fallacy

The best-known example of this fallacy was originated by Amos Tversky and Daniel Kahneman. It goes something like this:

> Linda is 31 years old, single, outspoken, and very bright. She majored in philosophy. As a student, she was deeply concerned with issues of discrimination and social justice, and also participated in antinuclear demonstrations.

Now that you know a bit about Linda, please circle the sentence you think is a more probable description of Linda.

1. Linda is a bank teller.

2. Linda is a bank teller and is active in the feminist movement.

The majority of people pick option 2. However, from a statistical perspective, the probability of two different events occurring at the same time (that is, in conjunction) has to be equal to or less than the probability of either of them occurring alone. Just think about the probability of lightning hitting your house the exact second that you run out of toilet paper. Obviously, the probability of both of them occurring at the same moment is lower than the probability of either of them occurring separately.

What is the process that leads people to make this mistake? It is called the **representativeness heuristic**. Mostly, we don't think about probability, even when we are asked to (it's too difficult), so we just ask ourselves which description is a better fit for the information that we have about Linda. The first description doesn't seem to fit, but the second one, despite its lower likelihood, feels like a better fit. It seems to represent Linda to a higher degree. And what about a possible link among the conjunction fallacy, personality traits, and misbelieving? Robert Brotherton and Christopher French, as well as Neil Dagnall and his colleagues, found that some people have a higher susceptibility to the conjunction fallacy and that these same individuals are more likely to be misbelievers.

Illusory Correlations

Lots of things around us change, and when we observe that things change together, we naturally think that they are connected. Sometimes, however, things do not actually change together, but we think that they do. That's an **illusory correlation**. For example, we might think that younger people are more selfish or that we feel more energetic after taking vitamins, not because these things are true but because we pay a different level of attention and have different expectations when we see younger people and when we take vitamins. (Just to be clear, I am not claiming in any way that younger people are more selfish or that vitamins cause anyone to be more energetic.) The point is that we see patterns where none exist (yes, this topic has been mentioned many times in these pages, but it is important to understand illusory correlations not just on their own but also as a part of a general set of heuristics).

What about a possible link between illusory correlations and misbelieving? As Michael Shermer noted, those who are more susceptible to illusory correlations are also more likely to be misbelievers.

Hindsight Bias

The name of this bias explains it all. It is the feeling that *we knew it all along*. Imagine, for example, that I ask you to go back in time five years and answer a few questions based on your opinions at that time. Five years ago, did you think that Tesla was going to be more, or less, profitable than Ford? What did you think was the chance that in 2022 Russia would invade Ukraine? What did you think was the chance that in the next five years there would be a global pandemic that would bring most of the world to a standstill? Of course, we all know the real outcome of these events, and knowing the outcome makes us feel that we would have more or less gotten the answers correct even five years ago. This is **hindsight bias**. Hopefully, these examples also demonstrate how easy it is to fall prey to it.

I named my research center at Duke University "The Center for Advanced Hindsight." I picked this name because early in my career, I would often describe the results of my research and get the response that they were obvious and unsurprising. No one would tell me that to my face, but the tone of their statement reflected a sentiment of "Why are you spending your time studying something so obvious?" The name, for me, was a way to warn people about the dangers of hindsight. I also wanted to remind myself and my team about the risk of hindsight—the risk that in retrospect we will convince ourselves that we knew a particular fact about human behavior all along. Hindsight bias is a very basic bias and it is difficult to overcome, so I don't think that the naming of my center is a sufficient reminder, but I hope that it helps a little.

By the way, to fight hindsight bias in my presentations, I now have two tricks. The first is that I usually collect not only data about what people *do* in an experiment (let's say, for example, that I'm conducting an experiment to encourage savings and I'm testing what condition leads to higher savings) but also data about what people *predict* the results of the experiment will be (what another group of people expect will lead to higher savings). First I present the prediction results, and (thanks to hindsight bias) people say to themselves, "Yes, this was obvious to me." Then I show the real results, which are often different from what people predicted, and the people who just convinced themselves that they knew the (wrong) results all along can't reuse their hindsight bias, so they are surprised. When I don't have a predictions study of this sort, I use a different approach and I ask the audience to vote on what they think the results will be. Once someone has committed to the—often wrong—result by raising their hand, the battle against hindsight bias has been won—at least for that specific question.

What about a possible link between hindsight bias and misbelieving? As Shermer noted, those with higher susceptibility to hindsight bias are also more likely to be misbelievers.

There are, of course, many other such decision-making errors and heuristics, and my point here is not to make an exhaustive list. Rather, it's important to observe the types of decision-making biases that are connected to a greater likelihood of misbelieving. What are those types? They include a lower level of statistical reasoning (the conjunction fallacy); a higher tendency to observe patterns where none exists (illusory correlations); and a higher tendency to trust one's intuitions, in this case having the feeling that we knew something all along (hindsight bias). Hopefully, by thinking about these types of decisions, you can look at other decision-making heuristics and consider whether they, too, are likely to be connected with a higher tendency to tumble down the funnel of misbelief.

The Case for Narcissism

Narcissism is not represented by a specific score on a scale, above which people are narcissists and below which they are not. Like all other personality scales, it is a spectrum, where each additional point means that people are ever so slightly more narcissistic. Some people score low on the narcissism scale, and some people score high. You can probably think about your family and friends and use your knowledge about them to score each of them on a scale from 0 (not narcissistic at all) to 100 (extremely narcissistic). Go ahead, take a minute or two to do that for a few people in your circle.

Okay, are you done? What's odd is that most likely, you don't even know the exact definition of narcissism, but nevertheless I'm going to guess that you feel confident in your scoring.

Just to set things straight, here is a short definition of narcissistic personality disorder, based on the *Diagnostic and Statistical Manual of Mental Disorders*, 5th edition (*DSM-5*), the official psychological diagnosis manual:

> Has a pervasive pattern of grandiosity. Needs and requires excessive admiration. Has a grandiose sense of self-importance. Exaggerates achievements and talents. Expects to be recognized as superior without commensurate achievements. Is preoccupied with fantasies of unlimited success. Has a sense of entitlement. Lacks empathy. Takes advantage of others to achieve his or her own ends.

Reading this definition, you're probably thinking that it fits with the general idea you already had about narcissism. However, now that you have seen the definition, it might seem more serious and maybe you want to go back and reclassify your friends and family; hopefully you can lower the narcissism score you assigned them. But remember, we are talking about a spectrum, and the definition given above is the definition of the extreme disorder. Though the general term *narcissism* will be used throughout this section, it is important to note that we will not be discussing those who fall on the extreme end of the narcissism scale—in other words, pathology. Instead, we will discuss people who might score just slightly above the norm in terms of their narcissism. If you sorted all your friends and family on the narcissism spectrum, we would be discussing those who were a little higher than the rest. They might still be lovable, smart, generous, and so on, but they would nevertheless rate somewhat higher on the narcissism scale.

What does narcissism have to do with the process by which people become misbelievers? How could the two be related? When I thought about the misbelievers I know, I started to understand how narcissism and misbelief might go hand in hand. Maybe I am just making up stories about clouds, but here is how I have come to see the connection.

Imagine a narcissist who needs constant reinforcement of their own greatness. Our narcissist wakes up one morning and feels that something is not right. Maybe a war has broken out somewhere in the world and everyone is busy with that. Maybe there was a landing on the moon, and all anyone can talk about is what an amazing accomplishment it is for humanity. Maybe their significant other has lost their job and they take out some of their frustration on our narcissist. Maybe their kids are studying remotely and demanding every ounce of attention. With these things going on, our narcissist feels that they are just not getting the proper level of appreciation and admiration. Something feels very wrong, and it is obvious that nothing could be their fault (because . . . narcissism, remember?), so they look for someone to blame. As luck would have it, they find someone or something else to blame for the unacceptable level of attention and recognition they are receiving.

But the story doesn't end there. Stress has been mentioned multiple times already, and in the case of the funnel of misbelief, stress and narcissism create a dangerous mix. Stress, as I have said, has a negative effect on all of us, and it creates a deep need to find answers and regain a sense of control. In the case of narcissism, this need is especially pronounced. We can think of these two forces as amplifying each other such that high stress combined with high narcissism creates a very high level of discomfort. Stressed-out narcissists have a particularly strong need to explain what is going on in general, and specifically the lack of attention that they are receiving, and with that need comes a higher tendency to move down the funnel of misbelief.

HOPEFULLY HELPFUL

Tend to Your Narcissists

You know those people in your circle whom you scored as just a little above normally narcissistic? Give them a little extra love, affirmation, or encouragement; remind them of their place in your social circle and the part they play in the grander community; or just give them a hug. It might keep their frustrated need for attention from leading them astray.

In Summary . . .

Many of the mechanisms that draw people into the funnel of misbelief are common human characteristics. Yet not all human beings are equally susceptible—which makes sense, because we're not all the same. Our individual differences—our personalities—play a role in determining how susceptible we are to all the elements explored in these pages. Why are some people more susceptible than others? Individual differences and certain personality traits such as patternicity, the tendency to trust one's intuitions, certain decision-making biases, and narcissism add to the picture. With this in mind, let's turn to the final elements of the funnel, the ones that seal the deal: the social elements.

FIGURE 8

The personality elements of the funnel of misbelief

- Personality—broadly understood as individual differences—plays a role in explaining why some of us are more susceptible to misbelief than others.

- It is extremely difficult to do personality research on misbelievers, since they instinctively mistrust the motives of the researchers. However, some common traits have been observed.

- Being more prone to misremembering, falling into the trap of false recall and false recognition, feeds misbelief.

- Seeing patterns where none exist is linked to misbelief.

- Overtrusting our intuitions is linked to misbelief.

- Decision-making biases such as the conjunction fallacy, illusory correlations, and the hindsight bias are more pronounced in misbelievers.

- Narcissism plays a role in misbelief.

- Personality cannot be easily changed, but knowing which traits correlate with misbelief can help us to identify risky points.

PART V

THE SOCIAL ELEMENTS AND THE STORY OF TRIBALISM

CHAPTER 9

Ostracism, Belonging, and the Social Attraction of Misbelief

For there are two possible reactions to social ostracism—
either a man emerges determined to be better, purer, and
kindlier or he goes bad, challenges the world and does even
worse things. This last is by far the commonest reaction
to stigma.

—JOHN STEINBECK, *CANNERY ROW* (1945)

It goes without saying that as humans, one of our defining features is that we are inherently social. It is a deep part of our nature both as a species and as individuals. We move through the world as members of countless overlapping social circles, great and small: families; friend groups; schools; local communities; religious groups; sports teams; clubs; companies; political parties; nations; and of course, today, numerous online groups and social networks. Our identities, behaviors, and beliefs are shaped throughout our lifetimes by the people around us, so it is not surprising that when some of us change our beliefs and begin to move down the funnel of misbelief, social elements play an important part.

I'm presenting social elements as a fourth category for the sake of a somewhat orderly exploration of ideas. But in reality, the social elements interact with all the other elements and are often in play at every step that a person takes down the funnel of misbelief. In broad strokes, we can think about the social elements as responsible for three separate types of processes. The first is a process that attracts people to take their initial steps into the funnel of misbelief; the second is a process that keeps people engaged within the funnel of misbelief. And the third process, which we will discuss in the next chapter, accelerates people's descent, solidifies their commitment, and increases their involvement with the misbelieving subculture.

The Initial Social Attraction

The first part of the social mechanism, the attraction, is what draws people into the funnel of misbelief. This initial process is fueled by our deep need for a sense of social belonging and the blow that our psychological well-being takes when we don't feel that we belong. Did you ever feel shunned by the cool kids at school or worry about being picked last for sports teams? Did you ever find yourself doing something you didn't really want to do, such as smoking a cigarette or poking fun at a classmate, in order to feel accepted? As an adult, have you ever pretended to like your coworkers in the hope that they'll invite you out to happy hour? Or posted something on social media and felt bad when the only like was from your mom? If so, you're familiar with the all-too-human need to belong and the fear of social ostracism. It's a powerful force, as social science has revealed. We'll look at some studies that shed light on this in a moment, but first let's imagine how the opposing forces of belonging and ostracism might affect someone who is beginning their journey down the funnel of misbelief.

Imagine Chloe, a young woman who feels stressed and anxious about the state of the world and perhaps is facing difficulties in her personal life as well. She's always been suspicious of governments (with good reason, she will tell you), and she finds all the government-issued restrictions during the Covid-19 pandemic confusing and frustrating. Then one day she sees a video shared by a friend of a friend on Facebook suggesting that the government did not tell the public the whole truth about President John F. Kennedy's assassination. She is curious. Since she learned about Kennedy in school, she always thought that he seemed like a more enlightened leader than many presidents in her lifetime. She's drawn to his youth and idealism and feels sad that he didn't live to make more radical changes to the country. She watches more videos, questioning the narrative she was taught in school. She wasn't even born when the assassination occurred, but her parents have often described the unforgettable day from their own childhood. They, like many Americans, remember exactly where they were when they heard the terrible news. Now Chloe wonders if her parents and schoolteachers were just repeating the "official" version that was presented in the news—a version that concealed the truth. Maybe there was a plot by the CIA or members of the conservative establishment who didn't like Kennedy's more progressive ideas? She even mentions this theory a couple of times in front of friends and family. People laugh uncomfortably, and some start to gently tease her. "What's next, a tinfoil hat?" The level of ridicule by her friends and family might in fact be quite low, but she feels it intensely, much more intensely than they intend. Nobody is being mean to her or ignoring her calls, but she ever-so-slightly feels that she no longer fully belongs. Meanwhile, online, she's meeting people who share her thoughts and encourage her to pursue them—people who don't dismiss her concerns or laugh behind her back.

It is easy to see how in such a case Chloe would start redirecting her social energy. She might drift away from her family and friends and spend more time with the people who give her reinforcement for the things she is insecure about, the things that her family and friends are poking fun at her for. It would be understandable for her to be attracted to a group of people who share the beliefs she is just starting to form. This is where she will find the kind of social support that gives her a sense of belonging.

Chloe now finds herself at the center of a metaphorical tug-of-war, pulled from two sides. One end is held by her family and friends and the other by the group of misbelievers she has recently met. Before she started down the funnel of misbelief, the pull from her family and friends was very strong and the other side did not even exist. But as she starts her journey down the funnel (feeling stressed, feeling hard done by, being exposed to new and interesting misinformation), a rope appears between her and the misbelievers. At this point, due to the feeling of ostracism that she gets when her family and friends laugh at her or dismiss her, she feels that the side being pulled by her friends and family is substantially weakened. Her perception of rejection is most likely an overestimation, but it has a powerful effect nonetheless. She finds herself leaning toward the side that is less painful.

Her new friends make her feel smart, independent-minded, and antiestablishment. She starts engaging with them in conversations about witness tampering, the suspicious deaths, forged evidence, and the behavior of bullets. She never imagined she'd learn so much about history and ballistics! She feels she has gained a lot of knowledge that other people don't have and that she is one of the few who understand the truth. It's strangely compelling to her. Chloe's connection to the misbelievers strengthens and she leans on these social relationships as a way to compensate for the social support she feels she has lost from her family and friends. As her misbelief becomes more entrenched, her family and friends pull away even more. Now, Chloe's feelings are not exaggerated. At this point her friends and family really are frustrated and irritated by her fascination with the JFK story and offended by some of the material she shares online. They don't want to spend time with her and listen to her theories.

In our internet age, we can easily imagine how this process would unfold and how quickly Chloe could find herself in a new social universe. As her posts and opinions garner less and less support from her friends and family while getting more and more likes and supportive comments from misbelievers, she slowly changes. She diverts more of her attention to the new friends and more of her time to the activities that those circles socially reward. Though social media may have accelerated this process, the process itself is not a product of the internet. The same general process of shifting allegiance can also take place, albeit in a much weaker and slower way, in a world where relationships are conducted face-to-face. In another era, Chloe could have slowly made new acquaintances and been exposed to new ideas, but she would have been much less likely to start hanging out at a different bar or attending a different church from the one where her family and friends gather.

The Pain of Ostracism

The feeling of ostracism has been studied by social scientists, and the story of how this research came about is interesting on its own merits. It is also a reminder of how useful it is to pay close attention to our everyday life experiences and the way in which our environment can be a rich source of new insights.

One day, Kipling Williams, a social scientist, was walking his dog in the park. As he was strolling along, a Frisbee fell at his feet. He picked it up and tossed it back to one of the guys who'd been playing with it. Grinning, the guy tossed it back to him, and Williams found himself in the middle of an impromptu triangle toss. Back and forth they went a few times, but then the two friends who had been playing together before he stumbled across their path stopped throwing the Frisbee to him and went back to throwing it to each other. Williams felt bereft and excluded. He was surprised at how disappointed he felt as he resumed his walk with his dog. After all, he didn't know those people. They owed him nothing, and he was not even that interested in taking part in a long game of Frisbee. Still, he was hurt.

Inspired by his experience, Williams decided to study the feeling he'd felt that day in the park, which he identified as a sense of ostracism. He devised an experiment that would create a similar experience to what he had gone through. In his experiment, three people, two of whom were working with Williams, were told to wait for the experiment to begin. One of the two who were in the know picked up a ball that was sitting on the ground and began to toss it back and forth to the others, as if to pass the time. The third person did not know that this was actually part of the experiment. In some cases, that person was included in the game throughout, receiving the ball approximately one-third of the time. In other cases, after a few tosses, the unwitting participant was never thrown the ball again and was basically ignored in the same way that Williams had been ignored. At the end of the game, Williams pretended that the experiment was starting and he asked the participant to describe their feelings.

Williams found that even a short and not very meaningful experience of ostracism can lead to substantial emotional impacts in many aspects of life. Most notably, after an experience of being ostracized for a few minutes by people they did not know and did not particularly care about, participants reported increased feelings of sadness and anger, a reduction in self-esteem, belonging, and control, and even an increase in their level of overall distress.

Next, Williams wanted to study more than participants' self-reported feelings. He was curious about what physical changes might be observable in the brain during an experience of ostracism. The problem was, brain imaging requires subjects to be perfectly still; they can't be running around in a park tossing balls or Frisbees. So together with his colleagues Naomi Eisenberger and Matthew Lieberman, he came up with a virtual game called Cyberball that could be played while the participant was inside an fMRI machine. With that technology, the researchers could study the participant's brain activity while the participant was being ostracized.

Participants were told that they'd be taking part in a study involving the effects of mental visualization, and the Cyberball game was introduced as a way to exercise mental visualization skills. Participants were informed that they'd be playing with two or three other players. In fact, the "other players" were programmed into the computer, but the participants didn't know that. The experiment had three parts. For part 1, participants were told they would not be able to take part in the game because of technical difficulties, but they could watch the other "players." In part 2, the technical issues were apparently resolved, and participants were included in the game, receiving the ball as often as the others. In part 3, after a few virtual tosses, participants were excluded. Just as the Frisbee players in the park had done to Williams, the virtual players stopped throwing the ball to the participant.

The study showed that when people felt excluded from the game, the same areas in the brain lit up that one would expect to see when we experience physical pain. Social pain, it turns out, shares a common neuroanatomical basis with physical pain. These results suggest that the experience of being ostracized—even in small ways—is very central to the human experience and affects us quite negatively. This explains why Chloe feels even the mildest teasing from her family and friends so intensely and why she's so drawn to the misbelievers who don't dismiss or shun her.

The Nicest People Online—to Each Other

"Thank you for sharing!"

"Thank you for speaking out!"

"I appreciate your bravery and honesty."

"Yes!"

Heart emoji.

Hug emoji.

It's a uniquely disturbing and incongruous experience to read dozens and dozens of comments like these following a post suggesting that I should be tried for crimes against humanity and publicly executed. Yet even the most graphic and vitriolic posts about me (such as the one I shared in chapter 4 about the bull being led to slaughter, by Jon) were surrounded by an amazing feeling of love and social support—not toward me, of course, but toward the original poster and the other contributors. This is not unique to posts about me.

In general, the amount of appreciation and love for one another expressed in social media conversations among misbelievers is unprecedented in my experience. They exhibit an extraordinary degree of mutual affirmation in the forums I've visited, praising and congratulating one another effusively. Who are these kind, wonderful people who show so much support and admiration for one another, even as they talk about nefarious plots and evil, genocidal elites? From time to time, I would even see reactions where someone would be so excited about a newly revealed plot or villain that they would suggest that the person who had made the discovery should be awarded the Nobel Peace Prize or some other humanitarian honor. I can tell you that among academics, I have never seen anyone post about someone else deserving the Nobel Prize for anything, but among misbelievers, prizes and recognition are often suggested.

If you could read only the emotional tone of the commenters without the content, you would think that this is a collection of the kindest people on Earth. And then you might wonder what noble cause had brought so many wonderful people together. Maybe they are trying to end poverty? Or create equal access to education for every child on Earth? Or cure some awful disease? What you would never suspect is that the topics they are dealing with are various plots to end humanity as we know it.

How come they are so nice to one another, even as they talk about very dark and painful topics? I think it is because it is so clear to every one of them that they themselves and everyone around them need social support and approval. Over time, they have created a social norm of providing such support in abundance, to such an extent that every other social circle pales in comparison (with the possible exception of the first two weeks when two teenagers fall in love and just can't stop telling each other how inconceivably, incomprehensibly wonderful they find the other person).

HOPEFULLY HELPFUL

Start with Common Ground

In the social realm, the discussion or argument style of "us versus them" is inherently unproductive. When people are in their "soldier mindset" (as discussed in chapter 6), they don't listen carefully, and they constantly counterargue in their mind. My guess is that they finish making their mental counterarguments even before the other person has finished their sentence. And because we're all so good at making arguments to justify whatever position we hold, the conversation goes nowhere. This is why it's incredibly important to first establish commonalities, connections, shared interests, and so on, even with someone whose beliefs are very different from our own.

In his book *How Minds Change: The Surprising Science of Belief, Opinion, and Persuasion*, the science journalist David McRaney delved into some of the most effectively proven methods of changing minds, especially on controversial and loaded topics. Across all the methods he explored, which include deep canvassing (from chapter 3), Street Epistemology, and Smart Politics, the first step is always to establish rapport. In other words, make sure the other person knows that you are a friend, let them know that you are not here to shame them, and communicate that you are interested in hearing their story and having an open conversation about it. In the context of this part, this is a good way to make sure that whoever you are talking to, be it a misbeliever or not, does not feel ostracized and sees you as part of their in-group rather than as part of an out-group.

Finding a way to be in an in-group with a misbeliever doesn't mean you have to accept or agree with everything they say. It's okay to find some of their beliefs offensive or dangerous. But like one of those Venn diagrams where two circles overlap in the middle, you can find some point of intersection. The in-group/out-group framework is a particularly useful approach because in an in-group, we can talk about joint goals and joint struggles. To apply this approach, first pick a topic for which it's clear that you and the person you're talking with have joint things to work on—something local, perhaps, such as housing prices or increased crime. Or find a common human issue that features in both of your lives, such as raising teenagers. In this way, you will send the message that the ways in which we're the same are much larger than the ways in which we differ.

Social Maintenance

Once people become part of misbelievers' social groups, a different process kicks in: the process of maintaining their beliefs and sense of connection to the group. There are two major components of this process, one that gets people to solidify their misbeliefs and a second that gets people to solidify their social circles. In this part, we are going to discuss these two subcomponents of social maintenance— solidifying misbeliefs and solidifying social circles—as if they are fully separate mechanisms, but of course they are connected. And their connection is best illustrated by what Robert Cialdini originally called **social proof**. So let's take a moment to understand social proof.

Social Proof: Social Maintenance at Work

As we go about our lives, in many situations, we are not sure what the correct normative behavior is. So we look to other people for examples of it. In the most basic sense, that's what social proof is. Imagine you are invited for the first time to meet the king of England at Buckingham Palace and you are not sure how to behave; you would just look around and follow the herd: bow when they bow, move in the same direction and at the same pace as everyone else, and so on. The same is true, albeit to a lesser degree, if we join a new workplace, go to Burning Man for the first time, or attend a parent-teacher meeting at our kid's new school. In all of these cases, we look for social proof to tell us how to fit in. Social proof also controls our behaviors in online groups, especially groups to which we are newcomers and in which we are unsure about the prevailing norms.

By using the people we look up to as a source of inference for the right things to say, the right things to do, and the right things to believe in, social proof links our social circles with our behaviors and beliefs, particularly in cases of high uncertainty. One of my favorite papers on social proof comes from work by Jessica Nolan and her colleagues. In one of their studies, they asked Californians what would motivate them to save energy. Participants responded that they would be most motivated by the environmental importance of saving energy, followed by the benefits to society, followed by the possibility of saving money, and that the least important factor was whether other people were also saving energy or not. Note, that was what people *said* would motivate them—their naive psychology about themselves. But when the researchers measured what *actually* motivated people, they found that the most important factor was what other people were doing (social proof), while the remaining factors were less important. These results show that social proof is important, but because we don't see how it works on us, we don't intuitively appreciate its importance and end up underestimating it.

Now that we understand this important link between social circles and beliefs, we will ignore this link and talk about the two components separately.

How Social Groups Solidify Misbeliefs

How do our social groups push us deeper into misbelief? You may have heard about the classic Asch conformity experiments conducted in the 1950s in which participants looked at three lines on the right-hand side of a board and one line on the left-hand side of a board. The participants were asked to say which of the three lines on the right-hand side of the board (line A, B, or C) was the same length as the line on the left. What was so interesting about that experiment was that when the participants made their choices alone, they basically got the correct answer every time. But when they made their decision alongside a group of confederates (all of them working for Asch) and all those confederates confidently made the wrong choice (let's say that they said, "Line A," when the correct answer was line B), the real participant often also made the wrong choice. This is a classic example of the pressure we all feel to conform with the opinion of those around us.

Asch carried out many more versions of this basic experiment. In one of them he tested what would happen if the other people in the room (the confederates who worked for Asch) were high-social-status individuals. Since such experiments often take place at colleges, imagine that the confederates were upperclassmen who belonged to an elite fraternity or sorority. What do you think happened in those cases? The social pressure was even stronger and the number of decisions that were influenced by social pressure increased in line with the social desirability of belonging to that group.

The basic Asch experiment tells us that we're influenced by other people in general. We have a tendency to follow the crowd. What the "high-social-status" version of the Asch experiment shows is that when we look up to other people, we feel even more pressure to conform and agree with their opinions. Let's go back to Chloe for a moment. She's recently joined some social media groups, drawn in by the sense of belonging it made her feel, in contrast to the ostracism by her family and friends. She looks up to the people who act as leaders in the groups: the people who have been active in the groups for a long time, the ones other people clearly follow. Because she looks up to them, she is even likelier to feel the pressure to agree with them. This is how the process of solidifying misbeliefs operates: once people join a group, the social pressure leads to more agreement, which leads to more exposure, and the belief is maintained.

How Misbelief Solidifies Social Circles

If social circles solidify misbelief, the converse is true as well: misbelief solidifies social circles. I found a very clear (and sad) example of this process when I went to visit a couple named Stacy and Ron. Before Covid-19, Stacy was a "mainstream media" personality. Early in the pandemic, she took a detour and created her own internet morning show with Ron. On the new show, they questioned the logic of everything related to Covid-19—from the ways in which governments were working to the scientific establishment to the information and recommendations that were coming from the CDC and the World Health Organization. Initially, they asked mostly reasonable questions, but the show also gave a platform to a range of guests who were more inclined toward misbelief in their perspectives on both Covid-19 and life in general. The show did not explicitly come out in support of the opinions of those individuals. Instead, under the guise of freedom of speech and the innocent claim that "we are only asking questions," Stacy and Ron allowed those guests to say whatever they believed, confronted them with very little skepticism, and then had an open conversation to better understand their views and perspectives. Over time, the people who followed the show changed, and so did the content. Stacy and Ron's beliefs drifted with them.

I observed that process from afar with some curiosity, fascinated to see such a public example of two people descending into misbelief. About a year into the pandemic, I asked a mutual friend to introduce us. After a few phone calls, I went to visit Stacy and Ron at their home studio. The first thing they wanted to know was what I believed and did not believe about Covid-19, but I redirected the conversation back to them. I wanted to discuss what was going on in their social circles; specifically, how they were being treated by their pre-Covid friends and their families.

What they told me did not surprise me, but it was nevertheless sad to hear. It represented, I suspect, a general experience shared by many misbelievers. I learned that none of Stacy and Ron's pre-Covid friends had reached out to them to find out how they were doing and to offer support for the work they were doing by flying the antiestablishment flag. They also told me that some of their pre-Covid friends had stopped returning their phone calls and that most of them had dropped off their professional Facebook page. Some had even called them to accuse them of spreading fake news. Stacy told me that her sister had stopped talking to her altogether and that Ron's daughter-in-law was not willing to have him visit, so he had been unable to see his grandchildren in a while.

With basically everyone they had known before the pandemic departing their social circle, I wondered how their social life looked. They described in depth the many social exchanges they now have, both online and in their backyard with their new friends. What was the common thread uniting all those people? You guessed it: misbelief. In that way, misbelief solidified their new social circles. Their social gatherings now included recounting in detail the social shaming and pressure they were all experiencing and commiserating with one another about it all. They also talked with their new friends about the new government regulations that were surely coming; lamented the sorry state of affairs in the world; and strategized as to how they could get more people to see reality the way they saw it. It's understandable that they would focus almost exclusively on those topics, since they were what brought the group together and connected its members to one another. Sharing their misery most likely served both to create a support group and to bond them more deeply with one another, strengthening and consolidating their new social structure. This is how the process of solidifying the social structure operates once people have joined such a group. More exposure leads to more agreement, which leads to more exposure, and the social structure is maintained.

HOPEFULLY HELPFUL

Fight the Temptation to Ostracize

If you're wondering how we can stop people we know and love from tumbling down the funnel of misbelief, here's one important suggestion: try to not ostracize them, even subtly. After all, social ostracism starts with the people who are not misbelievers (nonmisbelievers?) turning away from the misbeliever. If misbelief has infected someone in your life, you are probably guilty of this understandable but unhelpful behavior. After all, it's not easy to have the same conversations again and again about ideas that seem obviously untrue and even offensive—not to mention that these conversations can turn family gatherings into ideological battlegrounds. Understandably, it can be embarrassing to invite that person to a party and find them sharing their theories with family, friends, or work colleagues. It's easier to just turn away, let their calls go to voicemail, stop inviting them over, ignore their social media posts, or even block them. But it's dangerous. Those of us with friends and family who start to slip into the funnel of misbelief have the power and responsibility to do something about it. We can recognize that even a small amount of social ostracism can have a powerful negative effect—likely greater than the people who are ostracizing realize or anticipate. Instead of ostracism, it is important to show social support, even if it feels difficult and uncomfortable to do so.

Why Was Ostracism Particularly Harsh During the Covid-19 Pandemic?

Ostracism hurts in any context—even in a contrived virtual ball game. But during the Covid-19 pandemic, people felt ostracized in unusually harsh ways. When, in early 2022, French president Emmanuel Macron publicly stated that his strategy was to "piss off" the people who refuse to get vaccinated against Covid-19, it was clear that the season of mass ostracism was in full bloom. And when he went on to call unvaccinated people "irresponsible" and declare that such people "are no longer citizens," the pain of the misbelievers was visible without the need for any fancy brain-imagining techniques. That level of ostracism was quite extreme, perhaps even unprecedented, at least in modern times. Can you imagine a world leader—or any normative person, for that matter—using such language to describe people with different religious, economic, or social beliefs? I can't (with perhaps the exception that some Brazilians hold similarly negative opinions about the moral fiber of Argentinian soccer fans—and the other way around).

What was it about Covid-19 misbelievers that provoked such strong sentiments? To answer this important question, let's take a short detour into one of my favorite experimental paradigms: the public goods game. I like the public goods game because it is elegant, but mostly because it points to one of the most important yet often overlooked building blocks of a functioning human society: working together!

Here is how the game works. Imagine that we are playing the game in New York City. We start by picking ten random people who live there. On the first day of the experiment, we call them in the morning and explain the rules of the game: "Congratulations, you have been selected as one of the ten players of this game. Your identity is safe with us. We will never tell anyone else that you are one of the players, and we will also never tell you who the other nine players are. (Sounds like the TV show *Squid Game*, I know, but trust me, it is not.) Every day, we will call you at nine a.m. and give you a hundred dollars. We will let you do one of two things with this hundred dollars: either you can keep the money for yourself, or you can put it into the public pot. If you take the money for yourself, we will transfer it immediately into your bank account. But if you elect to send the money to the public pot, we will immediately transfer the money to the pot and once all ten players have made their choices, the total amount in the pot will be multiplied by five. In the evening, all the money in the central pot will be equally divided among all ten players. The game will continue this way for a while."

How would this play out? On the first day, the ten people get the call, they are reminded of the rules, and they make their choice. The results show that they usually all decide to put their money into the central pot. Ten people each contribute $100, for a total of $1,000. The money is then multiplied by five, for a total of $5,000. In the evening, the $5,000 is divided among the ten participants, and each person gets $500. Life is good! Our ten players each received a gift of $100 in the morning, and go to sleep with $500. This continues for a while—until, at some point, one person decides to keep the daily $100 to themselves and give nothing to the public pot. What happens now? On that day, nine people put in their $100, making a total of $900. This amount is multiplied by five, making it $4,500, and in the evening the $4,500 is divided equally among the participants, including the selfish %&$@ who didn't contribute anything. Everybody gets $450, but the %&$@ ends up with $550—$100 from the morning, plus the $450 from the public pot.

Now let's deviate from the standard format of the public goods game to a twist that taps into a broader range of our psychology. We'll call it the public goods game with revenge. The game starts in the same way, as described above. But now imagine that you were one of the nine players who put their money into the public pot in the morning and when you get $450 in the evening, you realize that someone from the group kept their money to themselves and as a result you got less and the selfish %&$@ ended up with more than everyone else. How would you feel about that? What if you were given an option to pay $20 to set up a plan where the %&$@ who betrayed the group would lose $400? Would you do it? Would you take revenge? Maybe now, sitting comfortably on your sofa and just imagining the betrayal, you don't think you would, but my guess is that if it really happened, you would be willing to cut off your nose to spite your face and sacrifice $20 to punish that %&$@. By the way, to get in touch with the emotions of betrayal, just think about a serious dispute at work or a fight with an ex and, with that feeling fresh in your mind, answer the question again: Would you pay $20 to punish the %&$@ who just betrayed you?

Now that we've taken this slight detour, let's return to the question with which we began this section: Why was ostracism particularly harsh during Covid-19? Hopefully, the parallels between the pandemic years and the public goods game with revenge are quite clear. During Covid-19, the people who obeyed the rules and restrictions paid a social and financial price, while the people who did not adhere to the rules made things more difficult for everyone else. And in a pandemic, it is much worse because even a small number of people who do only what is best for themselves at any given moment, without considering the public good, can make it much worse and prolong everyone else's misery. It was that feeling—that the misbelievers were betraying the public good and potentially doing great damage—that made the people who played by the rules so angry with the misbelievers, leading to the urge to ostracize and punish them.

Unfortunately, as we've already discussed, ostracism and the desire to punish are counterproductive. It only pushed the misbelievers deeper into the funnel, fueled by powerful social forces.

An Unwelcome Gift That Keeps On Giving: The Effects of Ostracism on Others

As discussed, social ostracism has several negative effects on the person being excluded. Among them are depression, a more negative outlook on life, and the need to look for social and emotional support elsewhere. Sadly, the story does not end with the negative feelings felt by the person being ostracized; it also changes the ways in which that person acts toward other people. In this sense, ostracism is an unwelcome gift that keeps on giving.

As Jean Twenge and her colleagues have shown, being ostracized leads people to donate less money, volunteer less, and help less. They started their experiments by making some of their participants feel ostracized while others were not. They achieved that by telling some participants that due to their type of personality, they were likely to end up alone later in life; the majority of their friends and relationships would have drifted away by their midtwenties; if they got married, their marriages would be short lived; and in the long run they would find themselves alone. In contrast, they told the nonostracized participants, the ones in the control group, that due to their type of personality, they were likely to have rewarding relationships; the majority of their friends and relationships would last a lifetime; when they got married, their marriages would last a long time; and in the long run they would always have people who cared about them and were willing to help them out. At the end of that short process, there was one group of people that felt ostracized and one that felt perfectly fine.

At that point, they exposed the participants in both conditions to a task and measured how they behaved. For example, in one of the experiments, the experimenter mentioned that she was collecting money for the Student Emergency Fund. She added that it was a great cause and it would be wonderful if they donated. She also pointed out that it was totally fine if they didn't. There was a donation box on the table and the box had a sign on it that read "Student Emergency Fund: To assist undergraduates with unexpected expenses." Then the experimenter left the room.

As you might have guessed, the participants who had been made to feel ostracized were less likely to donate. In another experiment, after being made to feel ostracized or not, participants from both conditions, as if by chance, stumbled upon a person who had supposedly spilled a bunch of pencils on the floor. Did they help? Again, the participants who had been made to feel ostracized were less likely to help.

Those experiments demonstrate the downstream effects of ostracism. And the extra bad news is that those effects do not end with just negative emotions. They also create negative social behaviors. Kai-Tak Poon, Zhansheng Chen, and C. Nathan DeWall wanted to determine whether ostracism also increases the willingness to act dishonestly. For their experiment, they went back to the ostracism basics and first got the participants to play Cyberball. The participants were told that they were playing with two other participants for a total of thirty tosses. In fact, they were playing against a computer that randomly decided whether they would get about ten of the thirty tosses (the inclusion condition) or whether they would get only two of the first few tosses and then nothing for the rest of the game (the ostracism condition). At this point, you are most likely wondering what kind of evil geniuses are these people who study ostracism? But let me assure you, they are wonderful and warm people.

Next the participants were presented with fifteen anagrams (for example, dictionary = indicatory; dormitory = dirty room, editor = redo it), one at a time on a computer screen, for fifteen seconds each. Participants were given an envelope containing 30 Hong Kong dollars, and they were told that after each anagram there would be a short break during which they could take 2 Hong Kong dollars if they were sure they had correctly solved the anagram they'd just seen. What they did not know was that only eight of the fifteen anagrams were solvable, and seven were not. Again, as you can probably guess by now, feeling ostracized made the participants claim that they had solved more of the anagrams, including some of the nonsolvable anagrams. In short, they were less ethical and cheated more.

To sum up these findings: as people feel more ostracized, they lose some of their moral standards, both in terms of their social morals (viewing helping others as unimportant) and in terms of their personal morals (finding it more acceptable to take more money than they earned). I have to admit that these findings worry me to a large degree. If we translate these findings to the realm of social media, where people can so easily feel shunned and ostracized, we can begin to see why this might drive, for example, lower moral standards when it comes to sharing truthful information and showing kindness toward others. I've said it before, but it is worth saying again: As you see the funnel of misbelief pulling in people close to you, resist the temptation to ostracize. The cost is just too high. Focus on offering them social support, even with all the difficulties involved.

In this chapter, we've seen how the opposing social forces of ostracism and belonging simultaneously push and pull people into the funnel of misbelief. And we've seen how those same forces increase the chance that people will stay there. But misbelief is rarely a stable, unchanging state. It has a way of accelerating, drawing people deeper into the funnel and becoming more and more extreme. In the next chapter, we'll look at the role social forces play in this acceleration.

CHAPTER 10

The Social Accelerator

Brian: Look, you've got it all wrong! You don't need to follow me, you don't need to follow anybody! You've got to think for yourselves! You're all individuals!

The Crowd (in unison): Yes! We're all individuals!

Brian: You're all different!

The Crowd (in unison): Yes, we are all different!

Man in Crowd: I'm not.

—MONTY PYTHON'S LIFE OF BRIAN (1979)

How did *that* person come to believe *that* thing? This is the question that began our journey, and we've explored numerous ways in which an otherwise rational person can begin to believe ideas that seem highly irrational. But there are other questions that become interesting as we explore the social element of misbelief: Does that person *really* believe that thing? What drives them to adopt and even spread something they might not fully believe? Why might people share information that has been proven false? Why might they double down on beliefs that have been revealed to be misguided? Why do their beliefs become more and more extreme? To answer these questions we need to look at the power of social forces within a group and the ways in which humans seek to prove their loyalty, gain status, and maintain bonds.

Once people reach a certain point in the funnel, they are so deeply embedded in social networks of people who share their views that those networks and the social forces within them begin to play an outsized role in accelerating misbelief and making it very hard for them to escape the funnel. In this chapter, we'll focus on this final component of the social element: the way in which social forces accelerate misbelief.

As we turn our focus to social acceleration, we'll start with one of my favorite forces in psychology, an oldie but goodie that has shifted our understanding of human reasoning. I am talking about none other than **cognitive dissonance**. Although I'm sure you know at least something about cognitive dissonance, likely a lot, it is nevertheless worth revisiting the basic concept because of its important role in the funnel of misbelief and, in particular, as a social accelerator.

Cognitive Dissonance

One of the starting points for understanding cognitive dissonance was a study that Leon Festinger carried out with two of his close friends, Stanley Schachter and Henry Riecken. For this study, they took advantage of a somewhat unusual event that was unfolding in 1954 in Lake City, Minnesota (a pseudonym they used for the city for publication purposes).

Here is the event described in the *Lake City Herald*:

> ### Prophecy from Planet. Clarion Call to City: Flee That Flood. It'll Swamp Us on Dec. 21, Outer Space Tells Suburbanite
>
> Lake City will be destroyed by a flood from Great Lake just before dawn, Dec. 21, according to a suburban housewife. Mrs. Marian Keech, of 847 West School street, says the prophecy is not her own. It is the purport of many messages she has received by automatic writing, she says. . . . The messages, according to Mrs. Keech, are sent to her by superior beings from a planet called

"Clarion." These beings have been visiting the earth, she says, in what we call flying saucers. During their visits, she says, they have observed fault lines in the earth's crust that foretoken the deluge. Mrs. Keech reports she was told the flood will spread to form an inland sea stretching from the Arctic Circle to the Gulf of Mexico. At the same time, she says, a cataclysm will submerge the West Coast from Seattle, Wash., to Chile in South America.

Not many people would be drawn to Minnesota for their winter vacation, and even fewer people would be interested in spending it with people who believe that the world is ending soon thanks to superior beings from Clarion. But that was exactly the kind of event that Festinger, Riecken, and Schachter needed to explore their theory of cognitive dissonance, which is the discomfort we feel when our beliefs conflict with our behaviors, and the odd ways we resolve this discomfort.

How could that strange event supposedly happening in Minnesota help them understand cognitive dissonance? To start, they assumed that Lake City would *not* be destroyed by a flood coming from Great Lake just before dawn on December 21! And since Marian Keech had a lot of followers who believed her declarations, they wondered what would happen on December 22, when they faced the unexpected reality that, contrary to their beliefs, the world was unchanged.

Soon after the researchers arrived in Lake City, Marian Keech received another message that thanks to their activity and beliefs, she and her followers would not die in the flood. An extraterrestrial visitor would come to her house in a flying saucer just before the flood and take her and her followers to safety somewhere in outer space.

Now let's assume that Marian Keech had two types of followers—some with very weak belief in her predictions and some with very strong beliefs. How would you expect that the strength of a follower's belief in Marian Keech's predictions would impact their level of disappointment when they found that the world did not end? Which type would be more disappointed, more likely to abandon Marian Keech and return home?

The standard prediction is that those who were more committed would be more disappointed because they would experience a higher level of disappointment in Marian Keech and therefore be the first to leave. However, the cognitive dissonance prediction was the exact opposite. The researchers theorized that the more serious followers would have a higher need to justify their initial dedication to Marian Keech and that pressure would lead them to double down on their initial decision to follow her. That was exactly what they found. The strong believers, who experienced a higher level of dissonance, took extra steps to validate their original decision and their belief in Marian Keech. They even increased their efforts to convince more people that Marian Keech was the real deal.

How did cognitive dissonance work to produce such an outcome? In the case of the weak believers, they experienced a conflict between their supportive actions of Keech and the new information that she was not a true prophet. But because their belief was weak, that was not a real issue for them. They just accepted that their supportive actions had been a mistake and went on with their lives.

On the other hand, the strong believers had a much deeper conflict between their supportive actions of Keech and the new information that she was not a true prophet. Since they were strong believers and their actions had been more substantial (some of them had given her all their money, left their families to be with her, and so on), they could not simply admit that they had been wrong all along. Instead, their solution was to refuse to accept the new information. When our actions and new information don't fit, we feel intense discomfort and a desire for consistency, which drives us to distort whatever is easier to distort, which is often the new information. Following that approach, the strong believers rejected the new information and even increased their commitment to Keech, in order to reconvince themselves that they were correct from the start in all of their supportive actions of her.

You may be thinking that Mrs. Keech and her flying saucers are a fringe example. But you have likely encountered many cases of cognitive dissonance in your own life or seen them playing out in the lives of your friends. We've seen several in the pages of this book, including the example of Saul in chapter 7, who experienced such extreme cognitive dissonance when he caught the very disease he believed was a hoax that he created a new and more elaborate story to explain his illness, even as Covid-19 was killing him.

Much like other psychological forces, the basic idea of cognitive dissonance can be traced through the canon of literature and philosophy. In Aesop's fables from ancient Greece, the storyteller coined the term *sour grapes* to depict the opinion change of a fox who realized that he could not reach grapes that he wanted. After experiencing frustration, the fox solved his cognitive dissonance by convincing himself that the grapes must be sour and that in fact he was never interested in them.

Benjamin Franklin made a related observation but he called his version of cognitive dissonance an "old maxim." Franklin's basic idea was that "He that has once done you a kindness will be more ready to do you another, than he whom you yourself have obliged." In other words, once we have helped someone out, we tell ourselves that we must like that person, so we continue to be nice to that person rather than deal with the cognitive dissonance of perhaps feeling differently about them. In his autobiography, Franklin expands on this idea and explains how he dealt with the animosity of a rival legislator:

Having heard that he had in his library a certain very scarce and curious book, I wrote a note to him, expressing my desire of perusing that book, and requesting he would do me the favor of lending it to me for a few days. He sent it immediately, and I returne'd it in about a week with another note, expressing strongly my sense of the favor. When we next met in the House, he spoke to me (which he had never done before), and with great civility; and he ever after manifested a readiness to serve me on all occasions, so that we became great friends, and our friendship continued to his death.

In addition to helping us predict when people will be more or less likely to accept new information and modify their beliefs, cognitive dissonance sheds an important light on the age-old question of whether we run because we are afraid or we are afraid because we run. Generally, there are two ways to link behaviors and internal states. The first (and standard) view is that internal states cause behaviors (we run because we are afraid), and the second is that behaviors cause internal states (we are afraid because we run). Phrased in these terms, the finding of cognitive dissonance is that behaviors *can* cause internal states. Now, this does not mean that it is always the case that behavior drives internal states; it means that the influence can go both ways. Internal states cause behaviors and behaviors cause internal states under different circumstances.

When it comes to cognitive dissonance and misbelief, the connection should be clear: Actions (committing one's time and resources to a cause, for example) can lead to a change in opinions (believing more strongly in that cause, even if it turns out to be unreliable). Now, in the case of misbelievers, there are all kinds of actions, but the vast majority of the possible actions are social: talking to other people, protesting, posting unverified information, reacting online, arguing with those who don't agree, breaking ties with former friends and family, and so on. This is why cognitive dissonance fits so well within the social part of the funnel of misbelief and why it is a strong force in quickly pushing people down the funnel.

Fake Conspiracy, Real Community?

When Peter McIndoe looked out a window in Memphis and saw a sea of protesters and counterprotesters in the street below, he felt as if the world had descended into chaos. It was early 2017, shortly after the election of Donald Trump, and the protest was one of many women's marches that took place across the country that day (remember those pink furry hats with ears?). Feeling helpless, McIndoe was struck by a sudden impulse to respond to the nonsensical situation he saw with something equally nonsensical. He tore a poster off his wall, picked up a marker, and wrote three words on the back: "Birds Aren't Real." And then he went downstairs to join the counterprotesters.

Soon people started asking him about his sign, so he came up with a story: The birds we see flying around are really surveillance drones. The Deep State murdered all the real birds decades ago and replaced them with tiny, feathered robot-birds.

It was an absurd response to what felt like an absurd world— a moment of improvisational performance art. McIndoe had no plan when he wrote that sign, but as he found his groove and embellished his story, drawing on all the tropes of popular conspiracy theories, people gathered around him and someone started filming. Before he knew it, he'd gone viral and his joke had become a movement of sorts. McIndoe played along, creating a website, videos (complete with actors claiming to be former CIA bird-drone operators), merchandise, and billboards and becoming the face of a national "Bird Brigade." He maintained that persona for several years and was even interviewed on right-wing talk shows describing the avian genocide and the complex machinations of the CIA plot.

Some estimates say that hundreds of thousands of young people have joined this fake conspiracy. Most likely some of them (I hope most of them) know that our feathered friends are, in fact, flesh and blood but found a darkly humorous solace in the shared joke. Most interestingly, they found something real in the midst of the fiction: a sense of community. McIndoe described it to the *New York Times* as "a safe space for people to come together and process the conspiracy takeover of America. It's a way to laugh at the madness rather than be overcome by it."

I wish that we lived in a world where what McIndoe created would have been a safe and harmless community. But sadly, we don't. So perhaps it's not surprising that too many real misbelievers latched on to the "Birds Aren't Real" claim and McIndoe eventually felt compelled to come out of character. Now the misbelievers think he's a CIA spy.

Anyway, whatever the outcome of Birds Aren't Real, it's a fascinating social study. And by the way, have you noticed the way that pigeon over there is looking at you?

Using Extremity to Enhance Identity

Cognitive dissonance is not the only social acceleration mechanism. Another important and very regrettable mechanism is the use of extremity to enhance identity. Here's the basic idea: If we are part of group 2, how will people in group 2 know that we are loyal to the group? We could put a 2 tattoo on our body somewhere visible, or we could express a very extreme opinion that aligns with the group's ideology. Let's imagine a group of people who have decided to reduce their meat consumption for health, ethical, and environmental reasons. People in this group do their best to eat meat only occasionally. Now imagine a person who wants to show loyalty to the group and maybe even climb up the group's hierarchy and become one of its unofficial leaders. This person is afraid of needles, so getting a tattoo won't work. Just posting that they are abstaining from eating meat on a particular day will not make any noise or clear the path to a leadership position. They need to make a splash. Maybe they can declare that they detest all meat eaters, that they are also giving up all leatherwear, that they will no longer eat at any nonvegetarian restaurants, or that they will not attend any dinner parties where meat is served. Extreme opinions are a key to being noticed; to showing loyalty and commitment; and to climbing the real and imaginary social ladders within a group. This very sad mechanism makes it more understandable why extreme language is social currency on social media and why social media are such a trap of extreme opinions.

Of course, once extreme opinions are expressed, they tend to become the norm over time, and once they become the norm, people need to express themselves in even *more* extreme ways to stand out. This mechanism doesn't excuse verbal violence and hatred, but it does help us understand one component of its prevalence.

HOPEFULLY HELPFUL

Don't Take Extremity Too Literally

Understanding the social role that extreme opinions play in establishing identity and demonstrating loyalty can help us in situations where extreme opinions are being expressed by someone we know and love. When your best friend from college starts sharing her beliefs about how Jewish bankers are running the world, it's easy to take offense and stop answering her calls (especially if you're Jewish, but even if you're just an empathetic human who dislikes hateful stereotypes). Though you have every right to feel that way, as we've discussed, it's not terribly productive to stop talking to your friend, as it might just accelerate her descent into misbelief. The first time she expresses such opinions, you might recognize that she is in a dangerous position and slipping deeper into the funnel. Her words may have less to do with her actual beliefs and more to do with her enmeshment in a community of misbelievers. When we have the presence of mind to see beyond our own feelings of offense, we are sometimes able to lend a helping hand before the funnel pulls our friends further in.

The Sad Story of Nurse Nadine

To give you a concrete story about the process of escalating extremity, let me introduce you to another person I encountered in my explorations of the funnel of misbelief: Nurse Nadine. Nadine was (you guessed it) a nurse, who caught my attention not long after I started my journey into the world of the misbelievers. I followed her because she was an interesting and extreme character. If anyone came to her in her capacity as a nurse for the Covid-19 vaccination, she would recommend that they not go through with it. She was particularly forceful in her recommendation when it came to kids, telling parents about all the terrible side effects of these experimental potions and the irreversible damage they would cause their offspring. Whenever she managed to persuade parents, she would post about her success online and get a lot of support and affirmation from her social media buddies. Indeed, she was something of a hero in the misbeliever circles. That was in part because she did more than just talk; she took action. It was also because she was a medical professional and as such, she gave medical credence to their beliefs. Over time, Nurse Nadine expanded her activities and recruited other nurses to join her efforts to persuade more people, particularly parents, against taking the vaccine.

One day something changed. I noticed that Nurse Nadine was being attacked online by the misbelievers—the very people who had praised and encouraged her. What had happened? It turned out that somebody had found a photo of her with (gasp) the two-faced, half-bearded devil who also had a part-time gig in the Illuminati with Bill Gates. The picture had been taken after a lecture I had given about four years before Covid-19 struck. At the time, Nurse Nadine had come to my lecture, taken a picture with me, and posted it on Facebook. She must have forgotten all about it. When the misbelievers unearthed the picture and started resharing and commenting on it, her loyalty to the group was suddenly called into question. They speculated that she was a double agent working in cahoots with me in service of the establishment. How else could the picture be explained?

The accusations against Nurse Nadine grew rather quickly. One group of misbelievers proposed that her whole operation was a front and that her real purpose was to flush out all the medical professionals who opposed the vaccine and then hand the list over to the government. As further evidence, they presented an analysis of the frequency with which she had been suspended from all kinds of social networks. Their analysis showed that she was suspended less frequently than other people, which they interpreted as clear evidence of her involvement with the establishment.

Nurse Nadine had a lot riding on her social status within the group of misbelievers. How could she clear her name? She employed several strategies, including increasingly vicious attacks against me and a professionally produced video promoting her cause and her organization of nurses against the vaccine.

The misbelievers remained unconvinced. Their next response was to raise another question: Where had she gotten the money for the professional-looking video? Nurse Nadine defended herself, saying that everyone who helped her had worked for free, including the editor, the makeup artist, and the hairdresser. As all of that was developing, one of my instincts was to write Nurse Nadine a public message on her Facebook wall: "Nadine, we had a good run for a while, but let's publicly admit that we've been caught." I really wanted to see what would happen. I have observed a lot of misbelievers and knew how quickly they resort to infighting. I wondered if getting them to destroy one another was perhaps a good strategy. (If you are a fan of J.R.R. Tolkien's *The Lord of the Rings*, this may remind you of the Orcs, who sometimes start killing one another instead of their enemy.) If I could pit them against one another, I thought, and get them to divert their energy into their internal squabbles, maybe I would slow down their ability to further spread disinformation. Although I was curious about that strategy and just the thought amused me, I could not bring myself to actually do it. It just seemed that Nurse Nadine was already suffering too much.

A few days later, one of the widely acknowledged leaders of this group posted that she had met with Nurse Nadine and was willing to vouch for her intentions and actions. She asked people to trust her and drop the accusations, which they ended up doing. But Nurse Nadine never returned to her earlier level of activity or social prestige. Over the following months, her social media activity faded away. Watching that happen made me appreciate why Richard had been so hesitant to partner with me in my research or to speak out in my defense, even after he had admitted that he no longer thought I was part of the cabal.

Nurse Nadine's case is interesting because it illustrates the importance of the social hierarchy within online social groups and the role of extremism as a way to gain social capital and bind people more tightly to a group, all while the group as a whole becomes more extreme. It also illustrates how the fear of social ostracism continues to play a role. Except in such situations, misbelievers fear ostracism at the hands of their comrades, so they seek ever-more-extreme ways to prove that they belong. One final lesson from all of this: if you intend to join a group of misbelievers, make sure that you don't have any skeletons in your closet that would raise suspicion. You are dealing with misbelievers, after all.

Identity, Polarization, and the Acceleration of Misbelief

In some cases, those who express extreme views start believing the things they share even if their initial goal was only to increase their standing within a group. And then there are cases where the theories being shared are so outlandish or unlikely that we have to wonder: Do they *really* believe these things? If we were to sit the person down for a polygraph test and quiz them about whether they truly think the earth is flat, the grieving parents who lost their children to gun violence are just actors, or Hillary Clinton is a pedophile, what would we find? Would they (or the lie detector machine) reveal that perhaps their beliefs are not quite so literal? If so, why are they spreading such lies? Understanding the mechanics of social groups—especially those connected by shared beliefs, such as religious groups, sects, and cults—can help shed light on this question. As Jonathan Haidt suggested, the deliberate sharing of a lie can act as a **shibboleth**— a kind of linguistic password that identifies people within a group. "Many who study religion have noted that it's the very impossibility of a claim that makes it a good signal of one's commitment to the faith," he wrote. "You don't need faith to believe obvious things. Proclaiming that the election was stolen surely does play an identity-advertising role in today's America."

Taking that idea one step further, Michael Shermer suggests that the extreme and obviously questionable claims of some misbelievers serve as proxies for deeper truths that a particular group holds sacred, that they are willing to gloss over the surface inaccuracies because they believe in the deeper cause. For example, if you're part of a group that holds the right to bear arms as a powerful truth and is dedicated to protecting that right at any cost, you might be willing to adopt and give voice to the theory that mass shootings are staged by your opponents and overlook the ridiculous (and deeply offensive) nature of your claim. Or if you are suspicious of mainstream medicine and deeply concerned about the power of Big Pharma, it might not seem like such a big leap to say that Dr. Fauci (and Dan Ariely) are trying to kill us all by promoting vaccines.

All of this reinforces the deep entanglement between our beliefs and our social instincts and affiliations. This is especially true in an era of profound cultural and political polarization. When loyalty to a particular political persuasion becomes the overriding motive, facts become currencies of identity rather than objective truths.

A particularly strange example of this played out in the lead-up to the 2022 US midterm elections. At campaign events across the country, some Republican candidates railed against a bizarre and disturbing trend that was happening in schools: the teachers were putting out litter boxes for children who identify as cats. That's right, cats. As many as twenty different politicians mentioned the "growing crisis," including elected members of Congress, sparking a flurry of social media posts and a mention on a high-profile podcast. Before proceeding, I want to assure you that there is no truth to the claim, as confirmed by an extensive investigation by NBC News. But the way the claim spread proved instructive in understanding how social identity drives and accelerates misbelief.

It appears that the rumors began on social media among parents.

Most likely it started as an exaggeration intended to make a point about hot-button issues regarding gender politics in schools. It is easy to imagine one parent saying to another, "They're providing gender-neutral bathrooms now for kids who identify as nonbinary. What's next? Litter boxes for kids who identify as cats?" However it began, the rumor quickly spiraled out of control, until it was repeated as a fact. Like many misbeliefs, this story contains a grain of truth (recall the magnetofection theory from chapter 5). There is indeed a subculture of people who call themselves "Furries" and like to dress up as animals, but it is more a form of role play than an identity. And no, they don't use litter boxes. Furthermore, there is no evidence that this trend has taken hold in schools or that it is being accommodated by educators.

To some on the political Right, the accommodations being made in schools for gender-nonconforming students seem as disturbing and bizarre as the litter box story. And if the electorate isn't sufficiently shocked by what's going on, perhaps they need a more extreme example to rouse them out of their torpor. Did those politicians *really* believe the litter box story? Or did they just use it to galvanize voters in the fight against what they saw as the morally dangerous agenda of those promoting LGBTQ+ rights? Many claimed to have heard first-hand stories or seen videos; one celebrity podcast host even claimed to have been told about an instance by a friend's wife who had seen it with her own eyes. But not one claim has ever been substantiated. Yet the myth continues—a strange testament to the way in which polarization drives and accelerates misbelief.

Indeed, if you look at our partisan political landscape, you see the essence of the misbeliever's mindset. Everything is viewed through the lens of "How is the other side conspiring to take advantage? What is their hidden agenda?" In a sense, when we fall into reflexively partisan ways of looking at the world—and we all do so at one time or another—we're all to some extent misbelievers. We see a piece of information that comes from the other political side, and we immediately assume not only that it's wrong but that it's designed to deliberately hurt us and our side. A serious discussion of how to combat political polarization is beyond the scope of this book, but I hope that in our own lives, each of us can strive to rein in our tribalism and partisan impulses. I have no doubt that if more of us did so, we'd also make an impact on the issues of misinformation, misbelief, and social cohesion.

At Least a Word About Social Media

Throughout this book, the topic of social media has been largely ignored. Sure, we've discussed misbeliefs that have been floating around and echoed on social media platforms, but we have not explored in depth the different mechanisms that make social media so powerful, complex, and dangerous. And if you were hoping that I was finally getting to it, I'm sorry to disappoint you. This book is about the psychology of misbelief, not about the platforms and mechanisms that spread misinformation. However, I do want to mention one element of social networks as an illustration of the differences between our natural communication methods (the ones that have developed evolutionarily and that generally fit our human nature) and these new artificial forms of communication. It goes without saying that social networks were not developed to fit with natural human communication methods, and consequently, they have produced some very undesirable results. The good news, as you will see, is that it doesn't mean we should categorically abandon social networks. Indeed, if we took a different approach and envisioned social networks that would improve communication while keeping human nature in mind, we could create different versions of social networks that would provide wonderful tools for human flourishing. So here goes.

In the early days of my online infamy, I simply could not believe that social media platforms allowed this kind of behavior to continue unchecked. Initially, I tried to report all the hateful posts, hoping that the platforms would block the misbelievers, but time after time I got a reply saying that these hateful behaviors did not violate the accepted policies and community standards. It was rather surprising to me to learn that death threats and videos claiming that I was a bitter homicidal maniac were acceptable within their community standards, but that was the response I repeatedly received. At some point, I decided that I needed to escalate the question higher up the chain of command, so I placed calls to people in positions of power within one of the large social networks. Eventually, I managed to meet with the group that was responsible for curbing fake news.

After a few quick introductions, I told the team a bit about my general experience with the misbelievers on their platform and shared with them a few specific posts about me. My intention was to frame the discussion and also make them realize that real people end up suffering from what happens in the virtual space they manage. I quickly moved to discussing the larger topic: What might be a good way forward?

I asked them to set aside the exact way that the platform had been built and the features it does and doesn't have and instead to focus on what we know about good communication and consider what the platform *should* do in order to become a useful communication tool. The expressions on the faces around the (virtual) meeting room were somewhat confused, so I jumped right into my first example. Here's the gist of what I said.

Let's think about the ways that communications take place in the animal kingdom. One of the most basic communication principles is the **handicap principle**, which was originally described by the biologist Amotz Zahavi. What is the handicap principle? The classic example of the handicap principle is the peacock tail. A weak male peacock cannot survive with a heavy tail, because it increases the peacock's vulnerability to predators. On the other hand, a strong male peacock can survive with a heavy tail. This makes the heavy tail a good signal to share with the female peacock: "Look at me. I'm a very powerful peacock." Why is this a good signal? Because it has a cost. A male peacock that is weak is not going to live very long with a heavy tail, making the heavy tail a good indicator of the real strength of the male peacock. This is the handicap principle at work. The male peacock is providing a signal by having a handicap (a heavier tail) and in this way making sure that any imposters that are, in reality, weak males can't communicate that they are strong (well, they can, but it will not work out for them for long).

Basically, nature doesn't allow cheap talk. A peacock can't just write on his Facebook page that he is very strong. He has to sacrifice something to prove the statement in order to make sure that the communication is real. The handicap principle provides a way to foster honest communications and eliminate dishonest communications—maybe not perfectly honest, but close to it. The handicap principle is an extreme version of a broader category called **costly signals** (in general, the more costly a signal is, the more likely it is that the communication is honest). Imagine somebody who wants to communicate that he is wealthy. Saying "Hey, look at me, I am rich" could just be cheap talk. Wearing nice clothes and driving an expensive car might help, but these are low-cost signals since the items can be rented. If the person buys expensive gifts and shows that he owns a luxurious home, the signal becomes more costly and harder to fake, and people are more likely to believe it.

The handicap principle includes another interesting element, which is that communication starts with the receiver and not with the producer. This means that for a communication signal (heavy tail, luxurious home) to develop, it starts with choosing a signal that the receiver will believe. Let's imagine a male peacock with beautiful eyes and long eyelashes that wants to convince a female peacock that beautiful eyes and long eyelashes are what all the female peacocks should be into. From now on, all female peacocks should judge male peacocks by their beautiful eyes and long eyelashes. It won't work. Why? Because the communication has to start with what the receiver (the female in this case) will find credible, not with the signal that the producer wants to communicate.

A reliance on the handicap principle is obviously very different from the way social networks are designed. In general, social media platforms idealize free communication with no consequences, which sounds good until we realize where a platform that celebrates cheap talk could lead. On social media, people can signal whatever they want, with no relationship to facts. There is no handicap principle, and there is no selection of signals that the recipient would find useful. Cheap talk on social media is an easy way to communicate, and as time goes by, it has become the norm.

Social media platforms essentially violate the basic rules of natural communications and as such they are incompatible with the evolutionary instincts we have developed to treat and react to information. In our evolutionary history, we've learned to expect honest signals that are associated with a cost (the handicap principle), and as a consequence we have developed intuitive trust in the information we receive. After all, if the information is mostly accurate, trusting it is a good strategy. And then social networks emerged, and the cost of communications disappeared. The reality is that we can no longer trust all or even most of the information we get online, but we still have the same evolutionary instincts that lead us to intuitively trust the information even when it no longer deserves to be trusted.

The team I met with was very interested in the handicap principle idea and other basic building blocks of communication. On the one hand, that was good news. On the other, it made it clear that the people in charge of controlling fake news on one of the biggest social media platforms are not very knowledgable about communication. They wanted to meet again, and over the next few weeks we talked about other principles such as reputation, trust, the feeling of anonymity, the voice of the silent majority, information bubbles, the emergence of negative social norms, extremism, de-escalating reactions in the heat of the moment, and more. During each meeting, we took time to discuss how these principles might be integrated into a version of their platform that would fit better with the ways the human mind has evolved, to create a platform that would have more positive effects. For example, we looked at ways to add incentives for truth telling and ways to enact the handicap principle.

As you might expect, the ideas flowed and people around the table were excited—until someone brought up considerations regarding the platform's current financial model, and then the excitement tapered off. We stopped a few months later, when it was clear that although we were having very interesting discussions, they would not result in actual change.

Are Haters Less Intense in Real Life?

Critics of social media and the internet in general often claim that the relative anonymity and lack of physical proximity afforded by online spaces act as an accelerant to hate. It's easier to be hateful when we don't have to look each other in the face. No doubt there is some truth to this. Virtual spaces are perhaps governed by different social mores than physical spaces are. When I found myself the object of hate at the hands of the Covid deniers, I had an unexpected opportunity to test whether they would be nicer to me in real life than they were on Telegram.

It was a beautiful fall evening, and I met Oded, one of my best friends, for a beer at an outside bar. I had just returned from several days hiking, and my mind was more at peace than it had been in many months.

My calmness didn't last long. I heard the sound of a crowd chanting and marching. I looked up, and too late I realized that we had inadvertently chosen to sit in the direct path of a Covid deniers demonstration. They were close enough that I could read the placards they were carrying: "Don't touch our kids"; "Our health is not for sale"; "My body, my choice"; "Vaccine mandates are a step toward dictatorship"; "What was washed more in 2021: brains or hands?"

I lowered my head over my beer, acutely aware of my unusual, half-bearded face. Luckily, the protesters were marching on the other side of the street and there was a bus stuck in traffic between me and them. I hoped that the bus would not move until after they passed and would continue blocking me from their view. Maybe they wouldn't notice me. But two men and a woman broke off from the others and walked toward me. The woman stopped dead, pointed at me, and started screaming. "Murderer! Murderer! Psychopath!"

The guy joined in. "We've seen you! We know who you are! We know what you're doing! I've been suffering for two years because of *you!*"

I tried to respond, but he wasn't listening. The people sitting around me were shocked, and the manager of the café tried frantically to insert himself between me and the angry demonstrators. The altercation probably lasted less than two minutes, but it felt much longer. It reminded me of a motorcycle accident I had once been in. Everything moved in slow motion. Every shouted word and angry gesture heightened its clarity and intensity.

When they finally moved on, a chilling thought crossed my mind. I'd always told myself (and numerous experts agree) that the internet adds intensity to people's hate—that they'd never be quite so vociferous in person. But maybe the boundaries between the online and physical worlds are in fact not so clear, and maybe we should worry even more about online violence.

How High Social Stakes Make It Hard to Return from Misbelief

Sadly, it seems, the funnel of misbelief is more often than not a one-way street. Yes, I've heard stories of people who have come back, admitted the error of their ways, and mended their relationships with their family and friends. Many of their accounts have much in common with cult survivor stories. But there are far more stories of those who simply estranged themselves more deeply from their family, friends, and mainstream society at large, never to return. There are many reasons for this, as we've seen throughout this book, and again the social element plays an outsized role.

The famous quote attributed to Upton Sinclair, "It is difficult to get a man to understand something when his salary depends upon his not understanding it," sets the groundwork for understanding the difficulty of getting misbelievers to change their minds. This quote succinctly points out that we have a built-in bias to see reality from the perspective of our financial incentives and this bias is very difficult to override. Of course, financial motivations are a small part of our complex set of motivations. We can just as easily swap the word *salary* for "pride," "ego," "identity," or "belonging." The point is that the opinions we hold often fulfill all kinds of deep emotional, psychological, and social needs: a need to feel in control, a need to feel in the know, a need to feel loved and empowered, a need to belong. Therefore, the battle to change opinions is more complex than the battle to provide people with information that would lead them to abandon their previous opinions.

In this regard, social needs are particularly powerful. Like cults, communities of misbelievers exact a high price for perceived betrayals. If someone has already been ostracized by family and friends for their original embrace of misbelief, the psychological stakes of changing their beliefs again are very high. They risk losing their new

---

HOPEFULLY HELPFUL

Listen to Former Insiders

There are people who have "left" groups such as QAnon, and some of them have willingly shared their stories publicly and graciously explained their experience of indoctrination and their descent into misbelief. There is so much to learn from their experiences. I believe that people like this, as well as the friends and family members of former misbelievers, have a critical role to play in helping others climb out of the funnel and even preventing people from stepping over the brink in the first place. This is another reason why it is so important not to ostracize the misbelievers in our lives. If we do and they eventually find a way out of the funnel, they'll feel shame and resentment and be much less likely to speak openly about their experiences.

community, their new friends, and their new social networks. Perhaps they've gained status among these groups, like Nurse Nadine or Richard, and fear that potential loss. All of these social factors conspire to make it very difficult for people to come back from misbelief, just as it is difficult for people to leave cults or radicalized religious communities.

In Summary . . .

In this part, we've seen how powerful social forces are at work at every stage of a misbeliever's progress down the funnel. As you have probably noticed, my desire to separate the social element into three distinct components of initial attraction, maintenance, and acceleration is a bit artificial, but I hope that you nevertheless have found it a useful way to highlight the role of each specific mechanism involved. We've seen how even the slightest degree of ostracism by family and friends early in the journey can loom large in the consciousness of someone who is taking their first steps into misbelief, pushing them more forcefully into the funnel. At the same time, the sense of being welcomed, acknowledged, heard, respected, and praised by fellow misbelievers creates a powerful pull of belonging. Once the journey begins, misbelief is maintained by social forces as well. We've seen how beliefs reinforce community, and community reinforces beliefs. Finally, we have examined how social forces accelerate misbelief, driving people to embrace more extreme positions in order to prove loyalty and gain status. Fear of social rejection or ostracism keeps people embedded in those communities, making it very hard for them to change their beliefs.

FIGURE 9

The social elements of the funnel of misbelief

In the journey down the funnel of misbelief, social forces play a powerful role.

The sense of ostracism is a powerful driver of misbelief.

Social Attraction happens in the early stages, as a result of a "push and pull" dynamic, where we start feeling ostracized by our friends and family and pulled by the sense of belonging to a new community.

Social Maintenance happens once we are deeper into the funnel and embedded in a new social group. For example:

- We look for "social proof" of how to behave from the new group of people around us.

- The group solidifies our new beliefs.

- The new beliefs solidify our loyalty to the group as the sense of ostracism from our old social circles intensifies.

Social Acceleration occurs when we are deep in the funnel, and the social bonds with other misbelievers "seal the deal" and make it hard to escape. This happens in several ways:

- Cognitive dissonance causes people to double down on misbelief.

- The need to show loyalty and gain status drives people to greater extremes, and polarization accelerates this process.

- The fear of losing status and relationships within the new social group makes it hard for people to leave.

PART VI

MISBELIEF, TRUST, AND THE STORY OF OUR FUTURE

CHAPTER 11

Can We Afford to Trust Again— and Can We Afford Not To?

When we discover that someone we trusted can be trusted no longer, it forces us to reexamine the universe, to question the whole instinct and concept of trust.

—ADRIENNE RICH

"Think about it this way," said my misbelieving spirit guide, Richard. "Have you ever been betrayed by a lover?" He was trying to explain to me how he had lost his trust in so many people and institutions. He went on to describe the following scenario: Imagine that you learn that your partner has been having an affair for the past ten years. Once you find out that the person you love has been cheating on you, it erodes your trust in everything related to that person. You start looking back at the happy times you had together, and you wonder if any of it was real. Even if you stay together, you can't really trust them anymore. You look at your mutual friends with new suspicion, wondering if they knew all along. You start mistrusting everything about that person.

"This is the same way that we, the conspiracy theorists as you call us, feel about the media, the government, pharma, and the elites,"

Richard said. "One betrayal is all that it takes. This is why eventually we are going to win, because sooner or later everyone will have an experience of betrayal, and they will suddenly see the world as it really is."

His metaphor was clarifying. And it made me think more deeply about the topic of trust and how we lose it, on both an individual and a societal level. It's a topic we touched on in chapter 1, and it has been a subtext throughout this book, but before we part ways, trust deserves a more explicit discussion. From a larger perspective, all of this is a story about trust and the erosion of trust. In this chapter, we will discuss trust, try to understand it better, and frame improving it as an important objective for society. In this regard, the perspective of social science and behavioral economics can provide a very useful lens that sheds some light on where we are, where we're going, and how we can get there.

The Lubricant of Society

Sadly, in my view, the damaging effects of the funnel of misbelief are not just going to disappear; in fact, they are going to gain more strength and momentum. And this is where things get worrisome. Why? Because trust is one of the basic ingredients of a functioning society, and misbelief is chipping away at it and creating real risks for our ability to work together and overcome future obstacles together. If we stop to think about it, it is rather obvious that trust is an important force in the world. Maybe it is best to think about trust as the primary lubricant that makes the machinery of our society run efficiently.

Take money as an example. Our belief in money is largely about trust: trust in the banks that keep our money, trust that other people will view money in the same way we do, trust in the stock market, trust that the government will not print too much money and will keep its value stable. Insurance, of course, is all about trust. We hand over our money now, and we trust that the insurance company will pay if something bad happens to us (even though insurance companies are often not the most trusted organizations, they still very much depend on our trust). We trust our doctors, lawyers, and car mechanics. We trust the fifteen-year-old kid next door to babysit when we go on date night, and we trust a different teenager to date our teenager (maybe this is a case where the trust is somewhat limited). We trust yet another kid to feed the dog when we are on vacation and to take in our mail. We trust Amazon with our credit card information, and we even ask strangers at the airport to watch our bags when we run to the bathroom for five minutes (effectively telling them that if they want to steal our stuff, now is the time). We trust that the government will set safety standards for roads, bridges, and elevators, and then we trust corporations to comply with these standards. We trust in democracy, the police, the fire department, and the justice system. Even if our trust is incomplete or lower than we would like, we still have a lot of trust in many of our institutions.

As all these examples illustrate, trust plays a tremendous role in our lives. In fact, trust is like the proverbial story about fish not noticing that they are in water because they are surrounded by it all the time. Trust is so important that it is hard to imagine all the obstacles we would encounter if we had a lower level of trust. Sadly, I suspect that unless we deal directly with the need to increase trust, we will hit each of these obstacles sooner than we think.

Just think what would happen if we lost our trust in everything. We would stop believing in modern medicine and refuse lifesaving surgery. We would run to the bank, withdraw all our money, and convert it to cash and gold. We would arm ourselves, not trusting that the police would protect us or that the military is there to defend our borders. What about believing in the results of elections? And if we lost trust in the government, would we still pay our taxes? Would we obey the basic rules of society? If we did not trust the system, would we listen to requests from the government to save energy or to stay home when we feel sick? I hope you are starting to get the picture that the consequences of losing trust are deeply worrisome. Perhaps you can already see the effects of loss of trust playing out in people you know.

Though the importance of trust in our modern society is somewhat hidden from sight, it does play a crucial role. And the funnel of misbelief leads to an erosion of trust between people, trust in governments, and trust in our important institutions. Mistrust begets mistrust in a dangerous downward spiral.

How Does Mistrust Beget More Mistrust?

In my journey into the topic of misbelief, I witnessed many examples of how mistrust escalates. One such incident that stands out happened soon after the Covid-19 vaccines came out, when I became aware of the Vaccine Adverse Event Reporting System (VAERS), the US database that collects information from the public about the side effects of all vaccines. Basically, anyone can go into VAERS and input information about side effects that they experience after a vaccination. The hope is that this information, at the population level, will provide starting data to help health professionals figure out what vaccines might be associated with what side effects, study these further, and, when appropriate, improve the vaccine, change how it is administered, or even pull it off the market.

At the same time I learned about VAERS, I also learned about a different website that I was told had been created by parents whose kids had either died or become very ill as a result of the vaccine. The "parents' VAERS" was much more user friendly and made the information about side effects more easily accessible. Supposedly, the parents' website took its information from VAERS, so the content was supposed to be identical, just presented in a different and clearer way. However, in online discussions, misbelievers claimed that there were some reported cases that had once been on VAERS, still existed on the parents' VAERS, but had been deleted from VAERS. The online discussions made it clear that deleting such reports was not only bad on its own, it also created another reason to mistrust any information originating from the government. Richard sent me several alarmed messages about it. After all, if the government was deleting data about side effects from VAERS, what else might it be doing?

I decided to check for myself, and indeed there were cases that appeared on the parents' VAERS but not on VAERS. Logically speaking, such discrepancies could result either from cases being erased from VAERS (as the misbelievers suggested) or from cases being added to the parents' VAERS. Whatever the cause, there was certainly a reason to worry that something did not add up.

As luck would have it, a couple of days later I had dinner with a physician friend. We were unexpectedly joined by a third person who happened to work in the IT department at the FDA, which oversees various programs including VAERS. Of course, I felt a bit uncomfortable about asking that person about the erased cases, because there was no way to ask without at least suggesting that the FDA might be involved in hiding information. But I couldn't help myself.

To my surprise, the FDA IT person knew exactly what I was talking about and not only acknowledged the removal of some cases but said that it had been done under the supervision of their department. They explained that this was because foreign powers, mostly Russian and Iranian, had found a way to spread disinformation using VAERS. So when the FDA identified cases that had clearly come from such sources, it removed them from the system to keep the data accurate. If the parents' VAERS were to download information between the time that a fake report was originally added and the time that it was classified as disinformation and deleted, that would explain the mismatch. It sounded very reasonable to me, and we moved to the main course and other topics.

Later that night I reported my discovery to Richard and asked him what he thought about that interesting plot twist. He was hesitant to accept the possibility that the FDA was on the side of accurate information, but he also asked me why the people running VAERS didn't share that data issue with the rest of the world so that everyone would know and understand the discrepancies. Richard's point sounded reasonable to me, so the next day I wrote the FDA IT person and asked why the agency did not share the problem of data contamination publicly? The reply: it did not want to announce to the foreign powers that it was onto them. Okay, that made sense to me as well. I went back to Richard to share that information in the hope that his trust in VAERS would increase, maybe above the level of trust he had for the parents' VAERS. But he insisted that it was wrong of the agency to erase cases, that it was unclear what exactly it was erasing, and that the government could not be trusted. That was the end of that conversation.

That exchange made clear to me that once trust is eroded—once one starts doubting the FDA, for example—any action that contains even the slightest ambiguity can be interpreted as further evidence of wrongdoing and further grounds for mistrust.

HOPEFULLY HELPFUL

Practice Trust

Trust, in many ways, is like meditation. It is all about practice, and the more we practice trust, the better we get at it. This means that at times of extra stress and tension, increasing our trust in the people around us in ways that make them feel more trusted and more in control can have positive spillover effects. This isn't easy, especially when our instincts tell us to withdraw our trust and protect ourselves. Here are a few simple ways you can practice trust: Make yourself vulnerable and reveal more personal information to your close friends. Finalize agreements with a handshake, rather than lawyering up. When you go out to dinner with someone, pay for the meal and tell them they can pay next time. All of these might seem like small steps, and they are, but that is the point when practicing trust.

Sadly, the importance of trust becomes very clear only after we lose it. Still, I am hoping that as a society we will soon realize how important trust is and start working on improving it before much more damage is done.

Mistrust and the Media: A Question of Responsibility

My next step into the complex world of escalating mistrust came a few weeks after the VAERS incident. I was speaking with a doctor from a large health care organization, and we were discussing how physicians were overworked, particularly during the pandemic, and what the organization could do to try to reduce the burden. Before we ended our meeting, I couldn't resist asking her what she thought about all the online chatter about unreported vaccine side effects. To my surprise, she agreed that there was a problem. She said that she had observed a lot of side effects in her clinic that had not been reported and had been collecting such data from her patients. I asked her if she could share the data with me so that I could analyze them and maybe shed some objective light on the question. But she reminded me that they were medical data and she could share it only with government officials or in response to a formal request from one of the news channels.

Undaunted, I drove to the headquarters of a large newspaper and asked to meet with the editor in chief. I told him about the undocumented side effects and suggested that the newspaper should request and publish the doctor's data, both to expose the real story and by doing so maybe also win back some of the trust of the misbelievers. The editor told me he suspected that I was correct about the underreported side effects. However, he had no intention of publishing anything about them.

I was rather surprised to hear something like that from the editor of a newspaper, and my expression must have conveyed it, because he quickly went on to explain that he didn't want to publish anything on the underreported side effects because he suspected that the misbeliev-ers would use the published information in an unethical way and distort it so much that it would end up becoming disinformation. It was as if he was weighing the total amount of disinformation that would result if he published the story (which was a combination of adding truth in his newspaper and giving fuel for distortion to the misbelievers) against the amount of misinformation if he didn't publish the story.

I left the meeting feeling very confused and unsure about the re-sponsibility of newspapers. Where does the responsibility of a news-paper end? Does it end with its sharing the truth, or is it also its responsibility to make sure that the information it provides is in-terpretated accurately? And how much weight should it give to the chance that people will misuse and misinterpret its information? Is it its job to do this cost-benefit analysis for society as a whole, or is it its job to just print the truth? I was disappointed that he did not publish

The Motivation-Destroying Scourge of Bureaucracy

Before we move on from the topic of trust, indulge me for another moment as I explain the relationship between trust and one of the things I personally dislike most in the world: bureaucracy. Bureau-cracy is a system that evolves when we don't trust other people. In-stitutions create rules that reduce the chances that people will step out of line in one particular area, but they don't see that in the pro-cess of adding bureaucracy, they decrease the level of trust that the people they are trying to regulate will have in the system as a whole. When bureaucracy increases, people feel mistrusted and therefore have less goodwill toward the institutions (often governments) that are making the rules. Consequently, more things go wrong that trig-ger more rules, creating a negative cycle that increases mistrust. Bureaucracy may take care of the particular focal point of a given regulation, but it erodes overall goodwill and destroys people's mo-tivation to improve and contribute to the institution. Sadly, bureau-cracy is increasing almost everywhere.

the story, but I could see his point. I have to admit that I was happy I was not in his shoes, making the decisions he was having to make.

Wherever you land on that question, it is very clear from this story that distrust and misinformation feed on themselves. Once there is lower trust, the questions of what to share become more complex, and if people and organizations become fearful and strategic about what to share, very little hope exists for getting the truth out and rebuilding trust.

HOPEFULLY HELPFUL

Invite Trust by Demonstrating Care

To create trust, we need to demonstrate that we're willing to put others' interest above our own. Here's an example of this. Imagine that a server in a restaurant approaches a table of four and asks, "What can I get for you tonight?" The first guest says they want the fish. The server replies that the fish is not good today and he'd recommend the chicken. It's cheaper and better, he adds. Now imagine the same server with the same table, only in this instance the server recommends the steak over the fish. It is twice as expensive, he adds, but it's amazing. For each table, we measure how likely each diner is to take the server's recommendation about the food and how likely it is that each table will take his advice about the wine. As you might guess, the guests at the first table are much more likely to follow the server's advice. Why? Because the server was willing to sacrifice something (overall revenue from the meal and the size of the tip) from their own utility for the benefit of the guests (a costly signal, as we discussed in chapter 10). In the second case, the server might be giving good advice, but the guests can't tell if he's trying to help them or himself. If we want to create trust, we first need to show people in a tangible way that we care about them more than we care about ourselves.

Once a spiral of mistrust begins, someone needs to make the first move to break the cycle, which means extending trust rather than doubling down on mistrust. In our social relationships, the question of who should make the first move can be open for discussion. When a government is involved, I think the answer is clearer. Because of the asymmetrical power differential, it's incumbent upon the government to take the first steps.

The Temptation Not to Trust

On a personal level, I've always thought of myself as a trusting person. But a few years ago, I had a professional experience that caused me to question my inclination to trust. It involved a woman named Clare, a friend of a friend, whom I met and eventually agreed to collaborate with on a media project. She was smart, creative, and fun, and I looked forward to our working together. I even turned down another project in favor of working with her. I worked hard on my side of the project, producing a lot of video content, but when Clare edited the content, she added all kinds of special effects that I felt distorted the message. We couldn't agree on how to proceed; we ended up parting ways, and I agreed to pay for her time and the video editor's time, as well as a cancellation fee for the hosting site. That was when things got weird. To cut a long story short, I found out that she'd not been paying the editor nearly as much as she had told me, and there was no cancellation fee. In fact, we had been paid an advance, which she'd never shared with me. I was shocked at those financial betrayals, and when I challenged her, she was evasive and confusing. Eventually, I decided to simply drop the matter and move on.

Unfortunately, it was difficult for me to deal with the residue from that experience. We all have stories like this, in which someone betrays our trust. My guess is that most people share my first instinctive reaction: to review all the ways in which we tend to trust people and all the ways in which we might be susceptible to future betrayals.

As in Richard's example of the betrayed lover, one betrayal makes us rethink our trust more generally. Maybe we even think that we should never trust anyone again and will use a legal contract for every project. This resolution feels good. Our future seems more predictable and under control.

A few days later, I once again considered the topic of broken trust. I had indeed made a decision never to trust anyone again and to always use lawyers and contracts. But I thought that maybe before implementing that approach I should do a more deliberate cost-benefit analysis. After all, making good decisions is supposed to be my profession.

I asked myself: What have I gained over time by placing a lot of trust, maybe too much, in everyone I work with, and what have I lost? The losses were clear, mostly related to the Clare fiasco, a few financial episodes, and a few heartbreaks. But—and here is the interesting point—though the negatives of placing trust in people were very clear, the benefits and the upsides were elusive and somewhat hidden. I started by thinking about all the people working with me and the trust I placed in them (I usually tell the people working with me that I fully trust them, that I don't need any reports from them until the end of a project, but that I am at their service whenever they need my help). Clearly, I was getting a lot of benefits from my trust: We were managing many more projects with higher efficiency, people felt empowered, we were having more fun, we were not drowning under bureaucratic burdens, and so on. Though the benefits of trust were not as apparent and easy to measure, it became clear to me what would happen if I gave up on trust. Sure, I might protect myself from potential betrayal, but I would lose all those amazing benefits.

I decided that the benefits of trust dramatically outweigh the costs. As painful as the Clare experience had been, I needed to look at it as the cost of doing business and not throw the trust-baby out with the bathwater.

I think that this lesson is even more applicable on a societal scale. It can be tempting, when faced with a world in which trust is an ever-scarcer commodity, to withdraw our trust and look for ways to protect ourselves. Yet what we don't realize is that when we withdraw our trust, we create a larger ripple and reduce trust in the societal pool. When companies—or even worse, governments—do the same, they damage one of the most important fabrics that connects us to one another. Sure, withdrawing trust from a specific activity will probably reduce the likelihood of that particular abuse happening again. But the larger picture is that it will also likely lead to a reduction of overall trust and with it a reduction in goodwill and cooperation.

Every time we withdraw trust, it feels good in the short term, but it has a long-term negative effect. We can do amazing things together as groups and as a society, but our motivation to collaborate toward a common goal relies, to a large degree, on trust. The conundrum for individuals and institutions is this: How can we override our initial defensive instinct to withdraw our trust and instead extend our trust for the sake of the greater good? One of the biggest challenges we are facing as a society is a crisis of trust, and we have to face this crisis head-on if we are to stem the tide of misbelief.

CHAPTER 12

Why Superman Gives Me Hope

A Final Word (Not Really)

*Solutions nearly always come from the direction you least
expect, which means there's no point trying to look in that
direction because it won't be coming from there.*

—DOUGLAS ADAMS, *THE SALMON OF DOUBT*

"Dan, how are you spending your summer? What are you working
on?"

I got this question a lot from friends during the summer of
2022, and the conversation that followed always went something
like this:

Me: I'm trying to understand and write about Covid deniers, conspiracy theories, and more generally about the process of misbelieving, where people lose their trust in science, institutions,
governments, and the media.

Friend: Why are you wasting your time on this? Haven't you noticed
that Covid is over? There are no more Covid deniers around.

Me: I agree and disagree. To some degree the pandemic is over, but the impact that misbelief has on society is still very strong. Think about how difficult it would be for anyone who went down the path of misbelieving and became committed to an alternative narrative for such a long time to suddenly say, "Okay, I was wrong, sorry about my mistakes, let's move on." After such an investment of time and energy, the effects are not just personal, they also lead to deep changes in the misbelievers' social structure. Do you think that after all of this they would so easily move on?

Friend: But I don't see them anymore.

Me: Trust me, they are still around, and as suggested by cognitive dissonance, they are doubling down on and expanding their beliefs.

These déjà vu conversations made it clear to me that my friends did not understand the effectiveness of the funnel of misbelief: how deeply it changes people; how it compels people to a high level of commitment; and how difficult it is to escape. Now that we are at the end of this book, I hope that we will all have a better appreciation for the forces that create the funnel of misbelief. Emotional, cognitive, personality, and social elements each play a role, amplifying and re-inforcing one another as the process accelerates.

Does understanding all of this lead us to a grand solution to the problem of misbelief? Sadly, no, although I do believe it reveals many ways in which we can fight the funnel. Here's what I do hope this journey has done.

First, I hope that exploring in some depth the psychology of mis-belief has shed some light on the danger it brings with it and the extent of the phenomenon.

Second, I hope it is a helpful guide to some of the traps that may lie in our own path, making us a bit more careful as we decide what we believe and what we don't.

Third, I hope it will help us help those around us avoid some of the traps.

Fourth, and more generally, I hope that this book will also help us think about the complexity of human nature. On the one hand, there is no question that human nature is fantastic and fascinating. We can do extraordinary things. On the other, we can do tremendous damage to ourselves and our world.

Fifth, I hope this book can increase our appreciation for the importance of trust and for the implications of a lack of trust.

Why Superman Gives Me Hope

If you think too long about the structural and societal problems that drive the crisis of trust and spread of misinformation, it can be easy to feel a bit hopeless. Then you open your computer and read the latest headlines about the new AI technology, which can churn out misinformation that looks authentic and credible and can be designed to fit each of our individual personalities. Spreading falsehoods is cheaper, easier, and more scalable than ever before. Will we ever get it under control? Sure, some tech companies are working to create better guardrails and to improve our ability to spot fake news, but it's a bit like a game of whack-a-mole that we can't win. This is why we need to pay more attention to the human side of the problem—to understanding the inner journey of misbelief and work to mitigate it in ourselves, our loved ones, and our social circles. That's where I find hope—in human beings. Which brings me to my favorite superhuman.

Think for a moment about Superman. More specifically, think about Superman's skills compared to the skills of regular human beings. Imagine we were to create a table with one column on the left for Superman's skills and one column on the right for the corresponding human skills. In Superman's column, we would list items such as can fly, can see in the dark, can run very fast, can stand for a long time, can withstand heat, can withstand freezing cold, can hear from across great distances, can remember everything, and is able to ignore his phone while driving. In the corresponding place in the human skills column we would write no, no, no, no, no, no, no, no, and not so much.

This comparison between Superman and regular humans might be somewhat discouraging, but there is another way to look at it. Over the past three hundred years or so, humanity has gotten much closer to Superman. If you look not at our natural innate skills but at what we are able to do, it is clear that we have made tremendous strides. We are now able to cross great distances in the air, to move very fast on land, and to sit for many hours in our comfortable chairs and sofas. We have invented air-conditioning and fans, and we have heating systems and warm cloths. We have microphones, loudspeakers, phones, and videoconferences. We even have applications that can help us remember everything. As for the texting and driving, we're not there yet.

If you think about this list, but also about the differences in the ways Superman and humanity got to our respective performances, we clearly did not arrive at our current capabilities by improving on our natural abilities. In fact, in terms of our individual physical abilities we are probably becoming a bit slower and a bit weaker. We got closer to Superman's performance by creating an envelope of technology around us.

In essence, even with our frail physical bodies, our technology gives us Superman-like performance. Think about the chair you're probably sitting in right now. Somebody gave a lot of thought to the design of the back, the armrests, the wheels. A team of designers and engineers must have spent hundreds of hours thinking about what cushion would be the best fit for your behind, just so that you can sit comfortably for longer periods. Of course, our efforts to engineer our environment in a way that enhances our abilities extends beyond chairs. In the pursuit of superhuman abilities, we have invented all manner of things and continue to do so (just wait until we have personal drones that enable us to fly to the supermarket and back).

In the process of improving our physical environment, we have made our lives much more interesting, we have increased our life expectancy, but we have also made our lives much more complex. As a consequence of that increased complexity, we are now forced to tax our cognitive system to a much greater degree than human beings have ever done before. Think about our ancestors on the savanna. They did not have to make decisions about retirement, financial investments, where to send their kids to school, what media outlets to believe, and so on. Not to mention whether to buy cryptocurrency, how to choose among alternative medical treatments, and how to interpret the instruction manual for an IKEA dresser.

Don't get me wrong; modern life is incredible, and we should feel privileged every day that we live in these times. But we must also acknowledge that the modern environment has dramatically increased the complexity of the decisions we have to make day in and day out. And this sometimes leads us to make suboptimal decisions.

With this appreciation for human incompatibility with the complexity of our modern environment, what should we do? How should we move forward? This is where the Superman analogy is useful. In the same way that we don't ask people to become cold resistant and instead create sweaters and heaters, we should also not expect people to always make good decisions in the face of complex information. In the same way that we invented heaters and sweaters to help with our physical limitations, we should invent technologies that can help counteract our mental limitations. Sure, we can assume that people are perfectly rational and that every person will always make perfect decisions, but this assumption makes as much sense as assuming that people are physically perfect and Superman-like. If we acknowledge our limitations and our irrationalities, we could start building the equivalent of airplanes, bicycles, crutches, and pillows for our minds and thereby achieve a higher level of performance. As a final example, think about the car and how many features have been created to decrease our chances of making deadly mistakes. We have headlights to see in the dark. We have side mirrors because we are too lazy to turn our heads. We have blind-spot monitors for the parts the mirrors don't catch. We have lane detectors because our attention drifts every so often. We have speedometers to help us know how fast we are going. We have annoying beeps from the seat belt mechanism to make sure that we don't forget this important safety feature. Almost any feature that you can think about was born of trial and error (mostly error). People driving without these features made mistakes, killing or injuring themselves and others in the process, and the amazing engineers working for the car companies came up with new features to reduce our tendency to make these particular mistakes. Basically, those features protect us from ourselves.

The same approach is important for other complex areas of our lives (financial, medical, health, relationships, education, and so on), and it is also important for disinformation and trust. It is tempting to assume that if information is shared on websites, by news outlets, in scientific journals, by government sources, and on social media platforms, people will react rationally to the information. By now I hope you agree that this makes as much sense as assuming that people can see in the dark. Instead, we must deepen our understanding of our limitations and build tools to help us rather than work against us. We've done it many times before, in many areas of life, and we can do it in this area, too. This is why I am optimistic. It doesn't mean that the journey ahead will be easy, only that it is possible to make the environment around us more supportive of the ways our minds work and in this way achieve much better outcomes.

In Conclusion . . .

So where does all this leave me? As I said, from the general Superman-inspired perspective, I'm optimistic. But on a personal level, I'm still working my way through everything I've learned and experienced about human nature, misbelief, and trust. Spending so much time with misbelievers and being attacked so brutally almost every day for the last few years has certainly done a number on my well-being, resilience, and general optimism. If you'd asked me ten years ago whether mistrust and misinformation were big issues, I'd have shrugged them off as lower on the urgency scale. Now they are at the top of my list. Now I view misinformation not just as inaccuracies that can simply be fixed by providing accurate information, but as corrosive untruths that can change a person deeply, often to a point of no return. I see this as a real threat to our ability to come together and solve the large challenges that confront us.

At times the forces at play and the outsized role of the funnel of misbelief in our lives seem too large and formidable to take on, let alone overcome. But humanity has improved in so many ways and managed to surmount so many obstacles. Why not this one? And yes, when a new technology is invented, it is often first used in questionable ways, but then we learn and fix things. Granted, the funnel of misbelief is not a piece of technology but is, instead, a complex tangle of human processes and external forces and technologies. As such, it is going to be very complex to fix. It's much more complex than just fixing social media platforms. As we've seen throughout this book, there are many things we can do to mitigate the problem, and there are many more things we don't know what to do about—yet. Hopefully, recognizing the importance of trust and the devastating effects of the funnel of misbelief will encourage us to take important steps in the right direction. And hopefully, by understanding the deeper psychology that underlies the issue of misbelief, we can begin to find ways to bridge the gaps and work together. In the end, despite being demonized, I was able to humanize, understand to some degree, and empathize with the people who demonized me. And that alone gives me reason to be optimistic.

FIGURE 10

As a final reminder, the funnel of misbelief and its constituent elements
(emotional, cognitive, personality, and social)

Acknowledgments

I am writing these lines during Thanksgiving, which is a wonderful time to think about gratitude in general and about my specific gratitude for the help with this book.

One of the main superpowers that made it possible for me to manage this complex period and come out only slightly damaged was Liron Frumerman. Liron and I have been friends for a while, and when the attacks started piling on, she took it upon herself to become my human buffer. She monitored the social media scene and passed on to me only the things she thought I must know about. She also joined many online groups, went to demonstrations, and in the line of duty almost agreed to go on a date with one of the misbelievers. In all these ways, she was not only one of my main sources of resilience, she was also instrumental in helping me understand the world of misbelief. I could not have hoped for a better friend and colleague for this complex journey.

The experience of diving into the world I describe here was complex and messy, and Ellen Daly not only helped me polish the ideas and clarify them with grace and humor; she also helped me sort out the messy experiences I was going through. Looking back at it now, I realize that she was also acting as my therapist—so an extra thanks for this. I could not have imagined a better partner to go through this adventure; my only regret is that we did not get to spend more time working side by side in person. I also want to thank Talia Krohn, who first encouraged me to move forward with this book and gave me very good advice about how to approach it. Dana Kindler helped me with a lot of the background research for this book; Nathaniel Barr and

Ben Heller provided very useful feedback and thoughts; and Rotem Schwartz created the beautiful graphics. Deep thanks to all of you.

As usual, I want to thank my amazing friend and agent, Jim Levine, for being there for me in every aspect of my profession and personal life for so many years. Deep thanks also go to my editor, Matt Harper, and my many friends at HarperCollins.

A special thanks goes to the researchers included in this book, whose work has been insightful, informative, and also helpful to me as I was trying to make sense of this complex phenomenon. I'd also like to thank the talented team at the Center for Advanced Hindsight, not only for the studies included in this book but for keeping my life at Duke interesting and fun.

Outside of the context of this book, I want to thank the people who make my life magical on a daily basis: Megan Hogerty and Yifah Hermony. A lot of what I view as wonderful in my life is based on the privilege of having both of you in my life. My deepest thanks and love.

Finally, I want to thank all the people who gave me some of their time, whether it was to listen to my ideas, share with me their ideas and experiences, or help me understand their perspective on what was going on in the world. I especially want to thank the many mis-believers who so generously shared their views and perspectives with me and spent hours communicating with me and arguing with me (and sometimes blaming me). I appreciate your willingness to engage with me and help me understand this complex world we live in.

Irrationally yours,

References

Introduction
Based on:
Anandi Mani, Sendhil Mullainathan, Eldar Shafir & Jiaying Zhao, "Poverty Impedes Cognitive Function," *Science* (August 30, 2013).

Chapters 1 and 2
Based on:
Figure 1 is based on Adam Enders, Christina Farhart, Joanne Miller, Joseph Uscinski, Kyle Saunders & Hugo Drochon, "Are Republicans and Conservatives More Likely to Believe Conspiracy Theories?," *Political Behavior* (2022).

Jonathan Haidt, *The Righteous Mind: Why Good People Are Divided by Politics and Religion* (New York: Pantheon, 2012).

Misinformation to promote political agendas in 2017: Sam Levin, "Fake News for Liberals: Misinformation Starts to Lean Left Under Trump," *Guardian* (February 6, 2017).

Additional Reading
Michael Shermer, *Conspiracy: Why the Rational Believe the Irrational* (Baltimore: Johns Hopkins University Press, 2022).

Harry Frankfurt, *On Bullshit* (Princeton, NJ: Princeton University Press, 2005).

Mikey Biddlestone, Ricky Green, Aleksandra Cichocka, Karen Douglas & Robbie Sutton, "A Systematic Review and Meta-analytic Synthesis of the Motives Associated with Conspiracy Beliefs," PsyArXiv (2022).

Jan-Willem van Prooijen, "Psychological Benefits of Believing Conspiracy Theories," *Current Opinion in Psychology* (2022).

Karen Douglas & Robbie Sutton, "Why Conspiracy Theories Matter: A Social Psychological Analysis," *European Review of Social Psychology* (2018).

Daniel Sullivan, Mark Landau & Zachary Rothschild, "An Existential Function of Enemyship: Evidence That People Attribute Influence to Personal and Political Enemies to Compensate for Threats to Control," *Journal of Personality and Social Psychology* (2010).

Chapters 3 and 4
Based on:
Shira Hebel-Sela, Anna Stefaniak, Daan Vandermeulen, Eli Adler, Boaz Hameiri & Eran Halperin, "Are Societies in Conflict More Susceptible to Believe in COVID-19 Conspiracy Theories? A 66 Nation Study," *Peace and Conflict: Journal of Peace Psychology* (published online, 2022).

Donald Dutton & Arthur Aron, "Some Evidence for Heightened Sexual Attraction Under Conditions of High Anxiety," *Journal of Personality and Social Psychology* (1974).

Martin Seligman & Steven Maier, "Failure to Escape Traumatic Shock," *Journal of Experimental Psychology* (1967).

Anandi Mani, Sendhil Mullainathan, Eldar Shafir & Jiaying Zhao, "Poverty Impedes Cognitive Function," *Science* (2013).

Eileen Chou, Bidhan Parmar, & Adam Galinsky, "Economic Insecurity Increases Physical Pain," *Psychological Science* (2016).

Jon Jachimowicz, Salah Chafik, Sabeth Munrat, Jaideep Prabhu & Elke Weber, "Community Trust Reduces Myopic Decisions of Low-Income Individuals," *Proceedings of the National Academy of Sciences of the United States of America* (2017).

Jon Jachimowicz, Barnabas Szaszi, Marcel Lukas, David Smerdon, Jaideep Prabhu & Elke Weber, "Higher Economic Inequality Intensifies the Financial Hardship of People Living in Poverty by Fraying the Community Buffer," *Nature Human Behaviour* (2020).

Joshua Kalla & David Broockman, "Reducing Exclusionary Attitudes Through Interpersonal Conversation: Evidence from Three Field Experiments," *American Political Science Review* (2020).

Kurt Gray & Daniel Wegner, "The Sting of Intentional Pain," *Psychological Science* (2008).

Real Time with Bill Maher, HBO, Season 19, Episode 31 (October 22, 2021).

Clive Thompson, "QAnon Is like a Game—a Most Dangerous Game," *Wired* (September 22, 2020).

Reed Berkowitz, "QAnon Resembles the Games I Design. But for Believers, There Is No Winning," *Washington Post* (May 11, 2021).

Additional Reading

Karen Douglas, Robbie Sutton & Aleksandra Cichocka, "The Psychology of Conspiracy Theories," *Current Directions in Psychological Science* (2017).

Stephan Lewandowsky & John Cook, "The Conspiracy Theory Handbook," Skeptical Science (2020).

Michael Butter & Peter Knight, eds., *Routledge Handbook of Conspiracy Theories* (London: Routledge, 2020).

Chapters 5 and 6

Based on:

Andy Norman, *Mental Immunity: Infectious Ideas, Mind-Parasites, and the Search for a Better Way to Think* (New York: Harper Wave, 2021).

Ruth Appel, Jon Roozenbeek, Rebecca Rayburn-Reeves, Jonathan Corbin, Josh Compton & Sander van der Linden, "Psychological Inoculation Improves Resilience Against Misinformation on Social Media," *Science Advances* (2022).

About the Socratic Method: Leonard Nelson, "The Socratic Method," *Thinking: The Journal of Philosophy for Children* (1980).

Peter Wason, "Reasoning About a Rule," *Quarterly Journal of Experimental Psychology* (1968).

Peter Johnson-Laird & Peter Wason, eds., *Thinking: Readings in Cognitive Science* (Cambridge, UK: Cambridge University Press, 1977).

Julia Galef, *The Scout Mindset: Why Some People See Things Clearly and Others Don't* (New York: Portfolio/Penguin, 2021).

Donato Paolo Mancini, "Cheap Antiparasitic Could Cut Chance of Covid-19 Deaths by Up to 75%," *Financial Times* (January 20, 2021).

Jaimy Lee, "'You Will Not Believe What I've Just Found.' Inside the Ivermectin Saga: A Hacked Password, Mysterious Websites and Faulty Data," MarketWatch (February 7, 2022).

Troy Campbell & Aaron Kay, "Solution Aversion: On the Relation Between Ideology and Motivated Disbelief," *Journal of Personality and Social Psychology* (2014).

Rebecca Lawson, "The Science of Cycology: Failures to Understand How Everyday Objects Work," *Memory & Cognition* (2006).

Leonid Rozenblit & Frank Keil, "The Misunderstood Limits of Folk Science: An Illusion of Explanatory Depth," *Cognitive Science* (2002).

Ethan Meyers, Jeremy Gretton, Joshua Budge, Jonathan Fugelsang & Derek Koehler, "Broad Effects of Shallow Understanding: Explaining an Unrelated Phenomenon Exposes the Illusion of Explanatory Depth," working paper, Yale University (2023).

Benjamin Lyons, Jacob Montgomery, Andrew Guess, Brendan Nyhan & Jason Reifler, "Overconfidence in News Judgments Is Associated with False News Susceptibility," *Proceedings of the National Academy of Sciences of the United States of America* (2021).

Additional Reading

Petter Johansson, Lars Hall, Sverker Sikström & Andreas Olsson, "Failure to Detect Mismatches Between Intention and Outcome in a Simple Decision Task," *Science* (October 7, 2005).

Gordon Pennycook, James Allan Cheyne, Derek Koehler & Jonathan Fugelsang, "On the Belief That Beliefs Should Change According to Evidence: Implications for Conspiratorial, Moral, Paranormal, Political, Religious, and Science Beliefs," *Judgment and Decision Making* (2020).

Gordon Pennycook, Jabin Binnendyk & David Rand, "Overconfidently Conspiratorial: Conspiracy Believers Are Dispositionally Overconfident and Massively Overestimate How Much Others Agree with Them," working paper, University of Regina (2022).

Nicholas Light, Philip Fernbach, Nathaniel Rabb, Mugur Geana & Steven Sloman, "Knowledge Overconfidence Is Associated with Anti-Consensus Views on Controversial Scientific Issues," *Science Advances* (2022).

Chapters 7 and 8
Based on:

Susan Clancy, Richard McNally, Daniel Schacter, Mark Lenzenweger & Roger Pitman, "Memory Distortion in People Reporting Abduction by Aliens," *Journal of Abnormal Psychology* (2002).

"Sleep Paralysis: Symptoms, Causes, and Treatment," The Sleep Foundation (2022).

Michael Shermer, "Patternicity: Finding Meaningful Patterns in Meaningless Noise," *Scientific American* (December 1, 2008).

Jan-Willem van Prooijen, Karen Douglas & Clara De Inocencio, "Connecting the Dots: Illusory Pattern Perception Predicts Belief in Conspiracies and the Supernatural," *European Journal of Social Psychology* (2018).

Jennifer Whitson & Adam Galinsky, "Lacking Control Increases Illusory Pattern Perception," *Science* (2008).

Bronislaw Malinowski, *Magic, Science and Religion and Other Essays* (Long Grove, IL: Waveland Press, 1948).

P. V. Simonov, M. V Frolov, V. F. Evtushenko & E. P. Sviridov, "Effect of Emotional Stress on Recognition of Visual Patterns," *Aviation, Space, and Environmental Medicine* (1977).

George Gmelch, "Baseball Magic," *Trans-action* (1971).

The part on Trump and gut intuition is based on: Aaron Blake, "President Trump's Full Washington Post Interview Transcript, Annotated," *Washington Post* (November 27, 2018).

The Comprehensive Intellectual Humility Scale is taken from: Elizabeth Krumrei-Mancuso & Steven Rouse, "The Development and Validation of the Comprehensive Intellectual Humility Scale," *Journal of Personality Assessment* (2016).

Shauna Bowes, Thomas Costello, Winkie Ma & Scott Lilienfeld, "Looking Under the Tinfoil Hat: Clarifying the Personological and Psychopathological Correlates of Conspiracy Beliefs," *Journal of Personality* (2021).

Tenelle Porter & Karina Schumann, "Intellectual Humility and Openness to the Opposing View," *Self and Identity* (2018).

Shane Frederick, "Cognitive Reflection and Decision Making," *Journal of Economic Perspectives* (2005).

Gordon Pennycook & David Rand, "Lazy, Not Biased: Susceptibility to Partisan Fake News Is Better Explained by Lack of Reasoning Than by Motivated Reasoning," *Cognition* (2019).

Amos Tversky & Daniel Kahneman, "Extensional Versus Intuitive Reasoning: The Conjunction Fallacy in Probability Judgment," *Psychological Review* (1983).

Robert Brotherton & Christopher French, "Belief in Conspiracy Theories and Susceptibility to the Conjunction Fallacy," *Applied Cognitive Psychology* (2014).

Neil Dagnall, Andrew Denovan, Kenneth Drinkwater, Andrew Parker & Peter Clough, "Urban Legends and Paranormal Beliefs: The Role of Reality Testing and Schizotypy," *Frontiers in Psychology* (2017).

Steven Stroessner & Jason Plaks, "Illusory Correlation and Stereotype Formation: Tracing the Arc of Research over a Quarter Century," in *Cognitive Social Psychology: The Princeton Symposium on the Legacy and Future of Social Cognition*, edited by Gordon Moskowitz (Mahwah, NJ: Lawrence Erlbaum Associates, 2001).

Ziva Kunda, *Social Cognition: Making Sense of People* (Cambridge, MA: MIT Press, 1999).

Neal Roese & Kathleen Vohs, "Hindsight Bias," *Perspectives on Psychological Science* (2012).

Christian Jordan, Miranda Giacomin & Leia Kopp, "Let Go of Your (Inflated) Ego: Caring More About Others Reduces Narcissistic Tendencies," *Social and Personality Psychology Compass* (2014).

Additional Reading

Andreas Goreis & Martin Voracek, "A Systematic Review and Meta-Analysis of Psychological Research on Conspiracy Beliefs: Field Characteristics, Measurement Instruments, and Associations with Personality Traits," *Frontiers in Psychology* (2019).

Aleksandra Cichocka, Marta Marchlewska & Mikey Biddlestone, "Why Do Narcissists Find Conspiracy Theories So Appealing?," *Current Opinion in Psychology* (2022).

Chapters 9 and 10
Based on:
Kipling Williams, "Ostracism," *Annual Review of Psychology* (January 2007).
Naomi Eisenberger, Matthew Lieberman & Kipling Williams, "Does Rejection Hurt? An fMRI Study of Social Exclusion," *Science* (2003).
David McRaney, *How Minds Change: The Surprising Science of Belief, Opinion, and Persuasion* (New York: Portfolio/Penguin, 2022).
Robert Cialdini, *Influence: How and Why People Agree to Things* (New York: William Morrow, 1984).
Jessica Nolan, Paul Wesley Schultz, Robert Cialdini, Noah Goldstein & Vladas Griskevicius, "Normative Social Influence Is Underdetected," *Personality and Social Psychology Bulletin* (2008).
Solomon Asch, "Studies of Independence and Conformity. A Minority of One Against a Unanimous Majority," *Psychological Monographs: General and Applied* (1956).
Jean Twenge, Roy Baumeister, Nathan DeWall, Natalie Ciarocco & Michael Bartels, "Social Exclusion Decreases Prosocial Behavior," *Journal of Personality and Social Psychology* (2007).
Kai-Tak Poon, Zhansheng Chen & Nathan DeWall, "Feeling Entitled to More: Ostracism Increases Dishonest Behavior," *Personality and Social Psychology Bulletin* (2013).
Leon Festinger, Henry Riecken & Stanley Schachter, *When Prophecy Fails: A Social and Psychological Study of a Modern Group That Predicted the Destruction of the World* (New York: Harper & Row, 1964).
Taylor Lorenz, "Birds Aren't Real, or Are They? Inside a Gen Z Conspiracy Theory," *New York Times* (December 9, 2021).
Zoe Williams, "'The Lunacy Is Getting More Intense': How Birds Aren't Real Took On the Conspiracy Theorists," *Guardian* (April 14, 2022).
Jonathan Haidt, *The Righteous Mind: Why Good People Are Divided by Politics and Religion* (New York: Pantheon, 2012).
Tyler Kingkade, Ben Goggin, Ben Collins & Brandy Zadrozny, "How an Urban Myth About Litter Boxes in Schools Became a GOP Talking Point," NBC News (October 14, 2022).
Amotz Zahavi & Avishag Zahavi, *The Handicap Principle: A Missing Piece of Darwin's Puzzle* (London: Oxford University Press, 1997).
About Kipling Williams and ostracism: "Purdue Professor Studies the Pain of Ostracism," Purdue Today (January 13, 2011).
Additional Reading
Daniel Sullivan, Mark Landau & Zachary Rothschild, "An Existential Function of Enemyship: Evidence That People Attribute Influence to Personal and Political Enemies to Compensate for Threats to Control," *Journal of Personality and Social Psychology* (2010).
Antoine Marie & Michael Bang Petersen, "Political Conspiracy Theories as Tools for Mobilization and Signaling," *Current Opinion in Psychology* (2022).
Karen Douglas and Robbie Sutton, "What Are Conspiracy Theories? A Definitional

Approach to Their Correlates, Consequences, and Communication," *Annual Review of Psychology* (2023).

Anni Sternisko, Aleksandra Cichocka & Jay Van Bavel, "The Dark Side of Social Movements: Social Identity, Non-conformity, and the Lure of Conspiracy Theories," *Current Opinion in Psychology* (2020).

Zhiying Ren, Eugen Dimant & Maurice Schweitzer, "Beyond Belief: How Social Engagement Motives Influence the Spread of Conspiracy Theories," *Journal of Experimental Social Psychology* (2023).

Index

Note: Italic page numbers refer to figures.

About the Author

Dan Ariely is the James B. Duke Professor of Psychology and Behavioral Economics at Duke University. He is a founding member of the Center for Advanced Hindsight; a cocreator of the film documentary *(Dis)Honesty: The Truth About Lies*; and a three-time *New York Times* bestselling author. His books include *Predictably Irrational, The Upside of Irrationality, The (Honest) Truth About Dishonesty, Irrationally Yours, Payoff, Dollars and Sense,* and *Amazing Decisions.* His TED Talks have been viewed more than 27 million times.